Contents

The human genome underlies the fundamental unity of all members of the human family, as well as the recognition of their inherent dignity and diversity. In a symbolic sense, it is the heritage of humanity.

Article 1,
Universal Declaration on the Human Genome
and Human Rights,
Unesco, 1997

Foreword

by Walter Schwimmer
Secretary General of the Council of Europe

The acceleration of scientific and technological progress is opening up constantly changing prospects full of both hopes and fears. As this work shows, for instance, recent discoveries in genetics mean that it is now possible to modify human beings themselves. However, this raises ethical questions about the use and control of this new knowledge.

Against this background, there is a growing need for real democratic debate on the many challenges arising from scientific and technological progress. It is with a view to fostering such debate that the Council of Europe has decided to start this new collection, focusing on the ethical dimension, the first volume of which is devoted to the human genome.

The public in all our countries must be provided with the relevant information and be enabled to play an active part in the debate, if we wish to avoid reductionist simplifications and manipulation, which often lead to tragedies.

As Secretary General of the Council of Europe, I would underline that the launch of a collection on ethics seems only natural in an Organisation that aims to promote, throughout Europe, an ideal of humanity based on respect for individual rights and human dignity, as well as democratic values. Through the collection, the Council of Europe wishes to share its ideas with the public in a spirit of coherence and openness.

Contributors

Jean-François Mattei

Professor Jean-François Mattei is a professor of paediatrics and genetics, member of the French parliament for the Bouches-du-Rhône area and of the Council of Europe Parliamentary Assembly, and President of the Liberal Democracy and Independents parliamentary group. He was the instigator of the French laws on bioethics (1994) and adoption reforms (1996 and 2000). Recently elected to the Academy of Medicine, he takes an extremely keen interest in health issues, and has written a number of articles. He was rapporteur on the investigation into BSE in 1997 and recently launched a petition on the Internet against gene patenting.

Paul Billings

Dr Paul Billings is a co-founder of GeneSage and the Editor-in-Chief of *GeneLetter*. An expert in clinical genetics and the impact of genetic technologies on society, he first worked as a professor and researcher at Harvard and Stanford before going on to become a dot.com entrepreneur. He has published over 100 articles and book chapters in addition to his book, *DNA on Trial: genetic identification and criminal justice*.

Sophia Koliopoulos

Ms Sophia Koliopoulos is the Managing Editor of *GeneLetter*. She has conducted original research into molecular genetics at New York University. She has also taught biology and has a background in journalism.

Juan-Ramón Lacadena

Professor of Genetics and Head of the Genetics Department, Faculty of Biology in Complutense University, Madrid, Spain, Professor Juan-Ramón Lacadena is a founder member of the Spanish Society of Genetics, of which he has also been Secretary (1973-1985) and President (1985-1990). A member of the Scientific Committee of the International Society of Bioethics (SIBI, Spain), he has written prolifically, producing university textbooks, articles and about seventy publications on genomics.

Jean Dausset

Winner of the Nobel Prize for Medicine (1980), Professor Jean Dausset obtained a doctorate in medicine from the Faculty of Medicine in Paris. He was Assistant Professor, and subsequently Professor, of Immuno-Haematology at the Faculty of Medicine in Paris, before being appointed Professor of Experimental Medicine at the Collège de France in 1977. His scientific career has been devoted to studying the major histocompatibility complex in human beings, using skin grafts from volunteers to demonstrate its importance for transplantation. In 1984 he set up a research laboratory – the Centre d'Etude du Polymorphisme Humain (CEPH), now known as the Fondation Jean Dausset – which constructs the physical and genetic map of the human genome and aims to identify genes responsible for genetic illnesses.

Robert Manaranche

Professor Robert Manaranche studied as a neurobiologist and obtained a doctorate in science. He then worked as a lecturer at the University of Paris VII (Denis Diderot), while pursuing his own research at a CNRS neurobiology laboratory in Gif-sur-Yvette. He held the post of Scientific Director of the French Association against Myopathy (AFM) until 1991, and in 1990 helped set up the Généthon laboratory, becoming its President from 1995 to 1998. He is now employed as a scientific consultant to the President of the AFM, casting a scientific eye over developments in genetics and the potential forms of treatment deriving from such developments.

Mike Furness and Kenny Pollock

Mike Furness and Dr Kenny Pollock both work on the LifeExpress Lead program for Incyte Genomics, Cambridge, UK. Mike Furness heads the program and is also a director of pharmacogenomics, and Kenny Pollock is a project manager. Mike Furness has worked at the University of London School of Pharmacy, the National Institute for Medical Research and the Imperial Cancer Research Fund, PE-Applied Biosystems, and most recently at the Molecular Pharmacology group, Pfizer Central Research.

Kenny Pollock has a scientific background in pharmacology and cellular signal transduction having carried out post-doctoral research in Cambridge and for ICI pharmaceuticals, now AstraZeneca. Prior to Incyte Dr Pollock worked for RPR in Dagenham, now Aventis, realising preclinical drug-discovery projects.

Bartha Maria Knoppers

Professor Bartha Maria Knoppers, Canadian Research Chair in Law and Medicine, is Professor of Law at the Faculty of Law, University of Montreal, where she directs the Genetics and Society Project. She is also the Chair of the International Ethics Committee of HUGO (Human Genome Organisation).

Jens Reich

Professor Dr Jens Reich is doctor of medicine and professor of bioinformatics at the medical faculty (Charité) of Humboldt University in Berlin. He is also head of the department of genomic bioinformatics at the Max-Delbrück-Center of Molecular Medicine. His field of research includes analysing the human genome for application in cardiovascular diseases and establishing the relationship between genetic constitution and lifestyle in the development of chronic metabolic damage such as arteriosclerosis and diabetes. He has written for numerous magazines and journals and is involved in politics in Germany.

Introduction

by Professor Jean-François Mattei

The birth of human genetics

The second half of the twentieth century has seen the sudden emergence of human genetics at the heart of the scientific revolution. It has quickly established itself as a new and original discipline touching on the very essence of humanity. Yet genetics has been going for as long as couples have been noting that their child has his/her mother's eyes and his/her father's ears. Where the phenomenon of resemblance is concerned, the notion of certain traits being passed down from generation to generation caught on very quickly. For centuries the transmission of hereditary characteristics was linked to the bloodline and history is not short of examples to give this idea credence, one being the haemophilia that dogged Queen Victoria's dynasty to extinction. In the nineteenth century, genetics was awoken from its slumbers by the theories of Lamarck, the observations of Darwin and the work of Mendel.[1] At the beginning of the 20th century the laws of heredity were rediscovered by Morgan and his studies on *Drosophilia melanogaster,* better known as the fruit fly. So genetics was not born yesterday.

Human genetics is new

Nevertheless human genetics does appear to be quite new and only very recently became established as a discipline in its own right, including in the medical field. A number of elements have come together to make that possible.

First there was the emergence of new knowledge such as the structure of the DNA molecule in 1953, the number of human chromosomes in 1956, the description of the first chromosome-linked illnesses following the discovery of trisomy 21 in 1959, followed by technological progress in cytogenetics and molecular biology. It was during this period that the link between certain identified genes and the occurrence of genetic illnesses was clearly established.

1.
See *Appendix I.*

But at the same time there have also been sweeping changes in public health needs. Child mortality has been substantially reduced by progress in the treatment of infectious illnesses and newborn child care. The priority is now shifting towards the frequency of malformations, congenital disabilities and genetic illnesses which affect 3% of children at birth. Preventing and caring for genetic illnesses is becoming a new concern, and all the more so since the advent of contraception fulfils the hitherto prime preoccupation of couples to control the number of children they had. With the problem of offspring numbers now solved, parents are spontaneously turning their attention to a drive for quality. So here we have new elements reflecting new needs and a change of mentality coming together.

Human genetics is original

But human genetics also seems to be an original discipline. Leaving aside a few rare exceptions, it has not yet moved into the realms of therapy. Oddly, it is required to understand the past, explain the present and predict the future. In this respect, geneticists seem a little like modern-day stargazers who have replaced the crystal ball with studies of the gene sequence.

So what we have is a situation in which geneticists are often required to make predictions for beings who do not yet exist or for the onset of a hypothetical illness. They are quizzed over the state of health of a child not yet born or the inevitability of a foretold destiny. It is interesting to draw a parallel here with the saga of medical microbiology which began over a century ago. After the discovery of bacteria it took decades to develop vaccines and antibiotics. Similarly, following the discovery of genes, we probably have a century's worth of work on genetic medicine ahead of us and it will doubtless take us decades to learn to effectively prevent and cure genetic illnesses.

Finally, an even more striking feature of genetics is that it generally steps outside the sphere of individual medicine and focuses on couples, families and sometimes even entire populations, since genes are shared. It is no longer solely a matter of one-to-one dialogue between doctor and patient since it is gen-

erally a whole family that is affected by the presence of a per-
nicious gene. An imperative set by genetics is therefore to successfully strike the delicate balance between respect for the
privacy of individuals with a legitimate claim to the confidentiality of data concerning them, and the interest of society
which may also legitimately demand protection against a
recognised and predictable illness. On the one hand there is
the sacrosanct nature of medical confidentiality and on the
other the pressing obligation to assist a person in danger.

Genetics and humanity

Indeed, partly because it focuses on humans in their social
context, genetics touches upon the very foundation of humanity, first and foremost in its historical aspect, as witnessed by
efforts to derail history for political ends. For obvious reasons,
the association of genetics and politics stirs up unbearable
memories, prompts extreme misgivings and justifies endless
precautions. The reluctance of politicians to tackle these terri-
bly difficult problems head-on is understandable. "Genetics",
"eugenics", "genocide" are all words that inevitably raise the
question of our historical humanity, since they lie at the origin
of the notion of crimes against humanity.

But genetics also touches on humanity in its spiritual dimension. Most of the issues it raises concern both life and death, in
other words the finite nature of man.[1] Ever seeking to better
understand who he is, where he comes from and where he is
going, there is in every man, and woman, a secret desire to
achieve immortality. But the only answer to death found so far
is the child. Obviously, then, the problems of sterility and malformations are all obstacles on that path, where genetics, both
cause and remedy, stands at the crossroads of ultimate questions. Beyond physical and moral suffering, it is inevitably the
metaphysical dimension that takes the upper hand.

The Human Genome Project (HGP)[2]

It is in this context that the project of sequencing the human
genome in its entirety was born: a crazy enterprise that has

1.
Editor's note: for man
read humankind.

2.
See page 34.

been compared to the Apollo project in the conquest of space, a dream of wanting to decipher what has been called, with heavy reliance on metaphors, the great Book of Life or the "genetic programme", incidentally endorsing the misconception that life was a story written in advance and could obey a computer program. It is, without a shadow of doubt, these expressions that gave rise to a terrible misunderstanding – never yet eradicated – that perceived genetics as a new form of ideology. Happily, the facts are quite different, and the most recent results of human genome sequencing are rather reassuring, even though they leave a great many questions unanswered for the foreseeable future.

The identification of genes

First there was the gradual identification of genes, as geneticists patiently plotted the thread of DNA, from seaweed to *Homo sapiens,* and decrypted its language, demonstrating the universal nature of the genetic code. In this way it has proved the universal nature of the living world and given it unity. From now on, questions as to man's place in the universe and even more so as to man's place in the order of living things are posed differently. Behind the identification of genes, the question of the dividing line between human and non-human emerges with renewed impact, with the realisation that human and non-human gene sequences may be so close as to generate confusion. Is the human gene fundamentally different from a non-human gene?

Then, by striving to isolate genes, genetics has learnt to identify them and to detect those which are beneficial and those which are not so beneficial, pursuing another objective which raises two kinds of problem. On the one hand, these concern the relative importance that should be attached to genes in the light of their environment, since there would be a tendency, were one to think solely in genetic terms, to misjudge the real effects of the environment on the mutation of sequences and the modulation of gene expression. On the other hand, there are the aspects of selection and elimination with a view to keeping, preferably, the best genes and eliminating the less beneficial ones. Solely in genetic terms, there is much at stake, for if nature exerts selection pressure, could the selection of genes by man not, in turn, exert pressure on nature? Ecosystems are delicately balanced and the issue here becomes the whole of biodiversity versus bioselectivity.

Furthermore, the recent data on the almost complete sequencing of the human genome do raise a number of new questions. The fact that humans possess only some 30000 genes overturns certain formerly held conceptions. The complexity of humans has nothing to do with the number of their genes, barely more than those of a fly or an earthworm. Opening up before us are whole new fields of investigation in which to seek a better understanding of the origin of humankind's infinite complexity. This is the new area of genomics which looks at the interaction of genes, the respective roles of non-coding DNA, transcription mechanisms (transcriptomes) and proteins themselves (proteomes), and probably other phenomena that escape us at present.

The ownership of genes

The potential ownership of genes is bound to raise at least as many questions as their identification. The problems raised by the patentability of living matter are well known, and debate is far from over between different cultures split by their differences. There has been plentiful discussion on the distinction to be drawn between discovery and invention: "discovery" is what one becomes aware of having been previously unaware of its existence, whereas "invention" is derived from intelligence and innovation and enriches human achievements. To what extent can one take ownership of living matter which, by its very nature, already existed? Can there be an exclusive commercial right over an aspect of living matter, part of our common living heritage? Is living matter not jointly owned by humanity? Put that way, it seems unacceptable, especially as the confiscation of knowledge is also a manner of hijacking the future. It creates a new form of organised economic dependency and further widens the gulf of inequality separating industrialised countries and developing countries. On the other hand, once the three additional requirements of industrial property – namely newness, invention and industrial applications – are fulfilled, patentability becomes logical and even necessary.

The use of genes

The issue of the use of genes must first be calibrated by conventional benchmarks which are above all indicators of common sense, universally acknowledged rules: "yes" to the use of genes if it can cure people, "no" if it makes them ill. I am caricaturing this self-evident position for a reason, namely to nip in the bud the prejudices skulking behind expressions such as "genetic manipulation". The choice of words is never innocent and some would like public opinion to believe that humankind might give in to temptation to modify the very essence of life, not always with the healthiest intentions. No one can forget that man has never once refrained from testing his knowledge in the light of reality, even if it has meant giving up the experiment later on.

It is in this context, and against the flow, that the magic word "therapeutic" comes into play. No one complains when it is a matter of introducing genes from foreign cells to obtain therapeutic substances such as insulin, growth hormone and interferon, to name but a few. It is even with a certain admiration that people hail a new form of pharmaceutical industry. Nor is there much criticism when genes are introduced into plants or animals for therapeutic purposes. Vaccine-bearing plants or genetically modified animals used as sources of human albumin or coagulation factors, for example, are fully accepted, as long as the rules of animal ethics are respected, as is the genetic modification of pigs with a view to xenografting hearts in humans.

This publication

All the questions we have just looked at merit detailed analysis from different specialist viewpoints, so that we gain a clear view of the debate and fully appreciate the issues. Doctors, scientists, law specialists, philosophers and politicians all have their arguments, from different standpoints, as do citizens and associations active in given sectors of society, including associations of illness sufferers. In terms of ethics it is of fundamental importance to try to state the issues as objectively as possible so that common solutions, acceptable to our human societies and reflecting our idea of humankind and due respect for human dignity, may be proposed. It is this kind of approach that forms the basis of ethical questioning.

Various countries have already adopted legislative provisions laying down the main guiding principles in the different fields concerned with biomedical ethics. But national law is inadequate in a world where trade and competition are such that common rules are required. Certain international organisations tend by nature to emphasise an economic approach to these problems. While this approach might be legitimate, it is not enough, given the nature of the challenges which go far beyond economic aspects alone and encompass philosophical, moral and metaphysical considerations in particular, which are the driving force in man's search for meaning.

It is the role of organisations more strongly oriented to upholding human rights, such as Unesco or the Council of Europe, to contribute to the drawing up of texts of reference for a world seeking a harmonious marriage between scientific progress and human dignity. It is their whole purpose to provide regular reminders that progress must serve humankind and that human beings are not to be an instrument of progress. The aim of the Oviedo Convention on Human Rights and Biomedicine, drawn up under the auspices of the Council of Europe in 1997, was to lay down a principal reference framework for biomedical ethics.[1] Discussion is ongoing within the Council's Steering Committee on Bioethics (CDBI) with a view to preparing additional protocols on more specific problems such as human cloning, organ transplants or research on embryos. Recommendations were also recently adopted by the Council of Europe's Parliamentary Assembly to protect the human genome from commercial exploitation and establish a framework for its use.[2]

1.
The text of the Convention can be accessed at: http://book.coe.fr/conv/en/ui/frm/f164-e.htm

2.
See Recommendation 1512: http://stars.coe.int/ta/ta01/EREC 1512.htm

It is in this frame of mind that this publication has been designed to facilitate access, in a simple and educational manner, to a variety of key information concerning the human genome, so that each and everyone can make up their own minds and take an active role in the debate.

Dr Paul Billings begins by focusing on a series of fundamental questions: What is the human genome? Why sequence the genome? Within what limits can we exploit what we have learnt about the human genome? Professor Juan-Ramón Lacadena then reviews the step forward from genetics to genomics, the genetic nature and diversity of humankind and the elements that would make up an ethical code required for human genetics. Professor Jean Dausset traces the emergence of the concept of preventive medicine and its development in connection with progress in genetics, emphasising the ethical limits that must not be transgressed. Professor Robert Manaranche describes the hopes for, and initial successes of, gene therapy, analysing the different approaches possible for a new branch of medicine. Mike Furness and Dr Kenny Pollock deal with the industrial aspect of the exploitation of the human genome, particularly as regards genomics and the pharmaceutical industry with all its developments. Professor Bartha Maria Knoppers looks at the most controversial aspects of the human genome, as individual property or common heritage. Finally, Professor Jens Reich considers the frontiers of the human being.

I am sure that these contributions will enable readers to gauge the immense progress accomplished in the field of genetic science, the marvellous hope given to sufferers and their families, the countless problems still to be resolved and the ethical issues that challenge man in his humanity.

What is the human genome?

by Dr Paul Billings
and Ms Sophia Koliopoulos

Human genetics, like all studies of heredity, seeks to explain two phenomena. First, why generation after generation, we resemble each other and our ancestors and develop in remarkably similar ways. And second, why, despite our similarities, we differ in ways that can be transmitted to our offspring, can make us differentially susceptible to disease and can provide the variation needed for natural selection and evolution. Given that the modern study of human genetics is barely one hundred years old (since the rediscovery of Mendel's principles,[1] their application to human traits and families, and the coining of the term "gene"), the depth, sophistication and promise of our current knowledge base are rather breathtaking.

In this chapter, we will review the latest information about human genetics and summarise why the human genome is a topic of such immense public interest. We will answer several basic questions:

1. What is the human genome?
2. What is the difference between human genetics and genomics?
3. Why sequence the human genome?
4. What is comparative genomics?
5. What are the limitations in using what we have learned from the human genome?

The answers to these questions should provide the reader with a basic understanding of this topic and its importance to biology and medicine.

What is the human genome?

While the true origin of the term "genome" is debated, many believe that it comes from a contraction of "gene" (or genotype) and "chromosome" – the microscopic structures that house genes inside most of our cells. Though continuing research

1.
See *Appendix I.*

keeps changing its meaning, the most basic definition of a gene is a unit of heredity that can transmit information to cells through its biochemical constituents. A simple way to define a genome is that it is the "gene's home" – a place where all the genes reside.

Almost all human cells contain two different genomes. A small one is located in the energy producing organelles, the mito-chondria,[1] in many cells. Its information is passed to the next generation primarily via the female egg (ovum) and so can be used to track matrilineage. The complete genetic map and sequence of the human mitochondrial genome has been available for many years (Anderson et al. 1981).

The other, much larger human genome is located in the nucleus of every human cell. It is comprised of approximately 6 billion bases of DNA, approximately half derived from each biological parent. DNA (deoxyribonucleic acid) is the heredi-tary material of which genes are made and is comprised of a linear series of bases – adenine, guanine, cytosine and thymine – that form pairs with a second DNA strand. The specificity of the pairing allows the series of DNA bases (arranged in a dou-ble helix) to be faithfully replicated as the number of cells grows from one fertilised egg into the trillions that comprise the adult human body. Through the sequence or order of these base constituents, via a chemical intermediary similar to DNA called RNA (ribonucleic acid), DNA participates in the produc-tion of proteins that help confer structure and function to human cells, participate in their metabolism and are part of other essential features of human life.

The human genome appears to be comprised of between 30000 and 40000 genes (Venter et al. 2001; International Human Genome Sequencing Consortium 2001) dispersed on 23 pairs of chromosomes. The 3 billion bases that make up a complete single copy of all the DNA found in the chromo-somes provides information used to produce the proteome (all the human body's proteins) of at least a hundred thousand proteins (International Human Genome Sequencing Consor-tium 2001). The number of proteins and the types of complex interactions resulting in human biological functions are not

1.
See page 138.

simply encoded in the DNA or represented by individual genes; rather, the amounts and activities of proteins are influenced by factors including other proteins, pharmaceuticals, nutrition and temperature, as well as regulatory parts of the genome itself. The DNA sequence is therefore not a "blueprint" for human life *per se,* but rather, instructions to produce a set of simple proteins and other non-protein producing information that together provide a necessary backdrop and guide to normal human development and function.

Only about 1% of human DNA present in the genome is directly involved in encoding information needed for protein production via RNA (Venter et al. 2001; International Human Genome Sequencing Consortium 2001). Thus, 99% of all human DNA either has no currently known function or is involved in regulating the activity of coding DNA (cDNA). Recent data suggest that DNA sequences in the non-coding areas of the genome, the so-called "junk DNA", may recombine and move throughout the human genome, possibly in response to cellular or environmental stressors. Thus the human genome may itself be a type of intracellular organelle that responds to changes in the environment.

Two independent research teams have now almost accomplished the complete sequence of the human genome (Venter et al. 2001; International Human Genome Sequencing Consortium 2001). The remarkable similarity of their results arises partly because of common data sources but also reflects the high degree of accuracy of currently automated DNA-sequencing methods and computational strategies. The data indicates that on average, we differ in DNA sequence approximately every 1000 to 1500 bases. We are thus about 99.9% identical at the level of our DNA sequences, even if our families are unrelated and we come from differing ethnicities, cultures and geographically distant areas. Only relatives, and particularly twins, share more than 99.9% of their genomes. The remarkable identity between unrelated individuals surely provides an explanation for biological similarity of humans and all living things, now and over the course of history.

Nucleotide: fundamental chemical building block of DNA.

Polymorphism: difference in DNA sequences among individuals.

Genetic differences, or mutations, can occur both in the non-coding areas of DNA that are frequently comprised of highly repetitive strings of DNA bases, and in protein encoding sequences. Single changes in the sequence of bases that occur frequently in populations or individuals are called "single nucleotide* polymorphisms*", or SNPs (pronounced "Snips"). Some SNPs result in altered protein sequences; SNPs may also affect protein levels and/or activity, while others may have no observable biological effect. With fast, low cost and highly accurate methods of detecting SNPs, their presence can be used to help predict genetic influences on a variety of important human traits (phenotypes) including aspects of human disease initiation, development, prognosis or responsiveness to treatment. Larger insertions or deletions of genetic material (often highly repetitive sequences), have applications in determining identity in forensic and paternity settings (Alford and Caskey 1994).

What is the difference between human genetics and genomics?

Human genetics arose during the classical period of the study of genetics that began early in the twentieth century. Traits that differed among individuals (usually in experimental systems using fruit flies or bacteria) were identified and the variation was ascribed to genes if it was heritable (passed predictably from one generation to the next) and could be mediated by the product of genes – proteins. Human genetics is the study of the role of individual or groups of genes in the production of human traits. Many genes have been identified and characterised because mutations in them produce a disease-related risk or trait, while others seem to be involved in only normal development.

With progress in methods for manipulating DNA that followed the discovery of the biochemical structure of the hereditary material by Watson and Crick in 1953 (Watson and Crick 1953), a new approach to the study of human genetics was developed. Termed "reverse genetics", it sidestepped the need for initially characterising how a protein was involved in a human trait. Using this approach, scientists first identify the physical location of a gene that may be associated with a human trait. Then the DNA sequence of that gene is analysed and general features of its protein product inferred from the order of the DNA bases. Normal and aberrant forms of that protein are then studied in order to describe their roles in the production of the trait being studied.

Improvements in analytical tools, including DNA sequencers and genotyping techniques that allow characterisation of large numbers and types of genes in a single experiment, along with the growing consideration of genes in the context of the entire human genome, led to the birth of human genomics, the study of our entire set of genes and their interactions. Thus, while human genetics focused on single or small numbers of genes, human genomics allows a comprehensive analysis of the genetic component of the differences in phenotypes.*

Why sequence the human genome?

The major benefit of determining the sequence of the human genome is its value in facilitating further scientific research, thereby yielding insight into human biology and evolution, as well as clues to prevent, detect and fight disease. Investigators around the world can access, via both free public[1] and subscription-based private databases,[2] the DNA sequence of the human genome and, in many cases, information about the biological roles of specific regions of the genome. These developments are already helping scientists develop new medical applications. For example, by referring to these databases, a team of scientists can compare the genes expressed in tumours with those expressed in normal tissue, possibly identifying genes that predispose or participate in types of cancer.

Information about the DNA sequence of the human genome is also leading to new insights into the evolution of the human species and the history of its communities. For instance, more than 200 of the human genes sequenced appear to arise directly from bacterial counterparts (International Human Genome Sequencing Consortium 2001). This suggests that in addition to large duplications of its own content, the human

Phenotype: the physical properties produced by gene expression and environmental influences.

1. Internationally-accessible websites post human genome sequence data. These include the United States National Center for Biotechnology Information (http:// www.ncbi.nlm.nih.go v/entrez/query.fcgi? db=Genome), the University of California at Santa Cruz (http://genome. ucsc.edu/) and Celera Genomics (http:// public.celera.com/ index.cfm).

2. For-profit companies have developed annotated databases of the human genome for use by researchers via subscriptions by institutional customers. These include Celera Genomics (http:// www.celera.com/cds/ cds_tour_frameset. cfm) and Incyte Genomics (http:// www.incyte.com).

genome derived at least some of its basic information from DNA donated thousands of years ago by bacteria. Some experts even hypothesise that the integration and subsequent duplication of bacterial DNA into the human genome may have been essential to the creation of the first human.

Ultimately, the practical value for the general public of knowing the full DNA sequence of the human genome will be measured by new medical treatments and preventive protocols that may arise from facilitated research. Refinements in the fields of proteomics, the study of a genome's full set of proteins, and functional genomics, the study of how genomes participate in phenotype production and biological functions, will be required to gain the full benefit of the sequence of the human genome.

What is comparative genomics?

The human genome is not the only focus of current genomic analysis. Shortly after the announcement of the completion of the human genome sequence last June, the complete sequence of the fruit fly, *Drosophila melanogaster,* was published (Myers et al. 2000). The publication of the human genome sequence findings in February 2001 was accompanied by the announcement that an important milestone in sequencing the mouse genome had also been achieved.

To date, the genomes of dozens of creatures – including a variety of bacteria, yeast, plants and animals – have been, or are in the process of being, completely mapped and sequenced. Detailed analysis and comparison of these genomes with each other and their human counterparts reveal several important findings:

1. Many animal species (vertebrates or invertebrates) and plants have genomes that contain numbers of genes that may actually exceed those found in humans.

2. In addition to sharing vast stretches of similar genomic DNA sequences with humans and being particularly useful for scientific study because of their rapid generation time, many organisms also have a genomic organisation that is

similar to that of humans. Zebra fish provide a much-studied model of this type.

3. The mouse genome, as well as those of chimpanzees and other Great Apes, probably differ from the human by as few as 500 genes or less (see the chapter **At the frontiers of humanity,** page 117). Therefore, various aspects of human complexity that distinguish our species from near genetic relatives cannot be explained simply by genetic differences among species.

4. The observed similarities between genomes of phenotypically diverse species has fostered the new concept of "the minimal genome", or the minimal set of genes that would support particular forms of life (Koonin 2000). There are many overlaps of the minimal human genome with other species.

Genotype: the particular type and arrangement of genes in each organism.

What are the limitations in using what we have learned from the human genome?

First, as previously noted, the vast complexity of human biological phenomena cannot be explained by 30 000 to 40 000 genes. The study of proteomics (the study of the proteome, the repertoire of proteins in a cell), functional genomics and other topics will have to provide additional and necessary insight to explain many important aspects of human biology.

Second, in many cases, genes may explain only a small part of the observed range of changes that a trait can undergo. Cancers, mental disorders and other conditions, for example, are not frequently inherited, suggesting that rather than being a cause of the condition, certain genetic characteristics that may be associated with their development may actually be a response to the disorder. Elucidation of the genome's role in these and other traits may be interesting and helpful, but it will not be sufficient to explain the whole range of phenotypic expression or disease causation.

Third, limitations in what we have learned from the human genome include the still prohibitive cost of genomic analysis. While many believe that genotyping analyses* will behave in accordance with Moore's Law, which, at the onset of the "computer revolution", linked increases in microprocessor speed with falls in cost, that progress remains to be accomplished. Emerging technologies do seem capable of sequencing DNA faster, easier and cheaper – an absolute requirement if universal access to one's own genomic information is to become a reality. For widespread use of DNA sequencing to become a reality, new tools for collecting, analysing, linking and interpreting the association of DNA sequences with medical and bioinformatic databases will also be necessary. Greater expertise in human development and genetic medicine among not only scientists, but also healthcare practitioners and patients, will be required. Thus, in addition to increased speed and reduced costs of DNA analysis, progress in information technologies, genetic epidemiology and training in genetic medicine will have to be established in order for full genomic analysis to become a standard of healthcare.

Finally, the genome is both several levels removed from the mechanisms involved in phenotype production and appears to interact in a complex way with itself and the world in which it resides (that is, in the nucleus, in the cell, within a tissue or organ, in a body within an environment). While we have developed sophisticated technology to assess DNA changes, our tools for monitoring the impacts of other types of biological activity and their influence on phenotype are less well defined. The fact is that DNA sequences will generally be linked to phenotype by probabilistic statements rarely approaching certainty. Uncertainty and unassessed influences will remain in most analyses. In fact this uncertainty remains a representation of our adaptation to a demanding and complex world. While perhaps to the frustration of genetic researchers and those who wish much of human life to be determined in the human genome sequence, the uncertainty will be both the genome's lasting enigma and our hope for continued happiness in a rapidly changing and surprising world.

The study of the human genome has benefited enormously from the development of technologies that allow observation of its sequence, association of specific regions of the genome with specific traits and comparison with the genomes of other species. Now, similar ingenuity will be required to interpret this data, link it to the proteome and understand how all this data yields structure and function. Chance, low probability interactions and pure environmental determination will always be factors in human life.

But the study of human genetics and genomics has already yielded a framework for understanding basic aspects of human heredity. DNA faithfully replicates and transfers information involved in protein production. All humans are nearly genetically identical, suggesting that the overwhelming similarity of humans and human development has a genetic basis. In other words, human difference may be partially genetic but primarily is not.

Yet genomes do vary among humans, and proteomes may be even more unique. Thus, in addition to explaining why we are so similar, studying the human genome can also elucidate ways in which we are unique.

Those searching for the answers to the universal questions of life, why it exists, why we die and what makes life worth living, will find the study and information derived from the human genome interesting. But it is neither the Holy Grail nor the Book of Life; those searching for those texts must continue their trek.

References

Alford RL, Caskey CT. "DNA analysis in forensics, disease and animal/plant identification." (1994) *Curr Opin Biotechnol* 5 (1): pp. 29-33

Anderson S, Bankier AT, Barrell BG, de Bruijn MH, et al. "Sequence and organization of the human mitochondrial genome." (1981) *Nature* 290 (5806): pp. 457-65

International Human Genome Sequencing Consortium. "Initial sequencing and analysis of the human genome." (2001) *Nature* 409 (6822): pp. 860-921

Koonin EV. "How many genes can make a cell: The Minimal-Gene-Set Concept." (2000) *Annual Review of Genomics and Human Genetics* 1: pp. 99-116

Myers EW, Sutton GG, Delcher AL, Dew IM, et al. "A whole-genome assembly of *Drosophila*." (2000) *Science* 287 (5461): pp. 2196-2204

Orkin, S.H., "Reverse genetics and human disease", (1986), in *Cell,* 47 (6), pp. 845-50

Venter JC, Adams MD, Myers EW, Li PW, et al. "The sequence of the human genome." (2001) *Science* 291 (5507): pp.1304-51

Watson JD, Crick FH. "Molecular structure of nucleic acids: a structure for deoxyribose nucleic acid." (1953) *Nature* 248 (451): p. 765

An ethical code for human genetics[1]

by Professor Juan-Ramón Lacadena

From genetics to genomics

At the present time, genomics is probably at the centre of genetics and even at the centre of biology, as Lander and Weinberg (2000) have pointed out. In order to understand this better, it would be appropriate to begin with a short overview of the historical and conceptual development of genetics.

Genetics: its historical and conceptual development

The birth of a new science – genetics – which explains biological hereditary phenomena was to be conditioned by its capacity to provide an answer to the two following fundamental questions: What are the laws by which biological characteristics are transmitted from parents to their children? What is the physical base, that is to say, the substance, by which such hereditary characteristics are conserved and transmitted? Or, in other words, what is the molecular basis of inheritance?

The answer to the first question was known from the experiments of Gregor Johann Mendel,[2] which were made public in 1865 in two consecutive sessions (8 February and 8 March) of the Naturalists Society of Brünn, Moravia (today Brno, Czech Republic) and published at the end of the following year.

The answer to the second question is intimately related to the history of deoxyribonucleic acid (DNA). In 1928, a scientist called Griffiths had already identified the fact that a particular substance was involved in the phenomenon of bacterial transformation. However, the identification of the hereditary material as DNA did not occur until 1944, when Avery, MacLeod and McCarty identified this substance. Thus, the "birth" of genetics can be said to have lasted eighty years, as it started in 1865 with the work of Mendel and finished in 1944 with the identification of DNA as the hereditary material.

1.
The present work is based on previous publications by the author (Lacadena 1992; 1993; 1996; 1999).

2.
See *Appendix I*.

These two questions give rise to two different definitions or concepts of genetics: one, which considers it as "the science that studies inheritance and variation in living beings" (Bateson 1906)[1] and the other, which I put forward (Lacadena 1981) as "the science that studies the hereditary material at any level or in any dimension".

According to the second definition, the object of genetics is the study of genes and, therefore, the formal content of this science has to provide suitable answers to the following questions about genes: What are they? How are they organised and transmitted? How and when are they expressed? How do they change? What is their destination in space and time?

The study of genetics can be divided into eight chronological stages until the present day, in terms of the main research carried out, see **Table I**.

1.
The word "genetics" was first used in a letter of 18 April, 1905, by William Bateson to Adam Sedgwick (published in *William Bateson, Naturalist*, 1928, ed. Beatrice Bateson, Cambridge University Press, p. 93). However, it was at the Conference on Hybridisation and Plant Breeding (William Bateson, chairman) held in London in 1906 that Bateson officially proposed the name of "genetics" for the new science. The transactions of the meeting were consequently published in the *Report of the third international conference on genetics*, 1906 ed. Rev W. Wilks, printed for the Royal Horticultural Society by Spottiswoode.

Table I: Chronology of major research

1865, 1900 – 1940	Transmission genetics *(Mendelian genetics)*
1940 – 1960	Nature and properties of *hereditary material*
1960 – 1975	Mechanisms of *gene action* (code, regulation, development)
1975 – 1985	*New genetics* (based on technology of nucleic acids)
1985 – 1990	*Inverse genetics* (genetic analysis going from genotype to phenotype, from gene to characteristic)
1990 – 2000	*Transgenesis* (horizontal transfer of genes (transgenic plants and animals; human gene therapy))
1995 -2000	*Genomics* (molecular dissection of the genome of the organisms (from bacteria to human beings: the Human Genome Project))
1997 – 2000	*Cloning* in mammals by nuclear transfer techniques

The year 1944 represents a fundamental milestone in the history of genetics because in that year deoxyribonucleic acid (DNA) was identified as the molecular basis of inheritance.

From that moment on, the progress of genetics has been continuous and it is becoming ever more rapid, going from abstract Mendelian "hereditary factors" to tangible genes that could be manipulated: genes are fragments of DNA that can be identified and isolated from the whole molecular mass of DNA that constitutes the genome of an organism. They can be characterised (that is to say, the genetic message they carry can be deciphered), transferred from one cell to another and from one individual to another, regardless of whether these individuals belong to the same species. This is of course genetic manipulation, where the term "manipulate" means "operate with the hands or with any instrument" and is not necessarily meant in the other possible pejorative sense.

Endonuclease: any of a group of enzymes that degrade DNA or RNA molecules by breaking linkages within polynucleotide chains. Restriction endonucleases recognise specific sites (consisting of base sequences) in the DNA and in the RNA.

The basic consequences and applications that have come from the identification of DNA as hereditary material are so numerous and broad that they have brought about one of the biggest paradigm shifts in the history of science. It can be said that in the history of genetics there is a "before DNA" and an "after DNA". The discipline can be divided into two time spans of more or less equal length: from 1865 when Mendel made his experiments public and 1900 when Mendel's principles were rediscovered until 1944 (the "before DNA" period), and from 1944 to the present day (the "after DNA" period).

Polymerase chain reaction (PCR): a method for creating millions of copies of a particular segment of DNA. If a scientist needs to detect the presence of a very small amount of a particular DNA sequence, PCR can be used to amplify the amount of that sequence until there are enough copies available to be detected.

In the years from 1975 to 1985, molecular techniques of restriction (fragmentation), hybridisation, sequencing and amplification of DNA were developed. These allowed, respectively:

1. DNA molecules to be cut where the researcher wanted to use "enzymatic scissors" such as restriction endonucleases*;

2. localisation of specific genes hybridising labelled probes with their complementary sequences in the original DNA;

3. direct reading of genetic information (which can now be done using automatic sequencing techniques);

4. multiplying by millions of times the amount of DNA available from a minute sample using the technique known as "polymerase chain reaction" (PCR)*.

This technology of nucleic acids is what has allowed genes to be manipulated: that is, genetic manipulation which gave rise to the "new genetics", as it has been named by the Nobel laureate Daniel Nathans.

Many years ago, Fred Hoyle, an astronomer at the University of Cambridge, UK, prophesied that "within twenty years, physicists, who only manufacture inoffensive hydrogen bombs, would be working freely, while molecular biologists would be carrying out their research behind barbed wire". What Hoyle was predicting was the enormous power that genetics was going to have for manipulating genes. The potential of genetics is enormous and this makes society perceive genetics as an all-powerful science, considering DNA as a new philosopher's stone of biology, although some people, in view of the negative uses to which genetic techniques can be put, see the double helix of DNA as a "double-edged" molecule.

This discovery of DNA has not only influenced genetics in particular, but also biology in general, and even society as a whole. With the benefit of hindsight, it seems obvious to think that history and philosophy of science will tend to include the DNA Revolution as a fundamental milestone in the history of mankind, in the same way that the Agricultural Revolution and the Industrial Revolution were also milestones in the history of mankind. While the development of technology has brought mankind towards a technocracy, the DNA revolution is, in some senses, leading to a biocracy.

Genomics

The term "genomics" refers to the branch of genetics that deals with the molecular dissection of the genome of organisms; in other words, the total knowledge of the sequence of bases of its DNA. From 1995 onwards the development of genomics began to produce spectacular results with the deciphering of the entire sequence of genomes of simple organisms, such as bacteria (in this period more than fifty species), followed by more complex organisms (the yeast *Saccharomyces cerevisiae,* the nematode *Caenorhabditis elegans,* the insect *Drosophila melanogaster,* the plant *Arabidopsis thaliana* and others still in progress such as the mouse *Mus musculus,* etc.), to finish with the Human Genome Project (HGP) as the most complex example of genomics.

The Human Genome Project (HGP)

Within genomics, the Human Genome Project represents its biggest challenge. The human genome is made up of around 3 picograms (1 pg = 10^{-12} g) of DNA, equivalent to 3 billion base pairs.* The Human Genome Project in its simplest terms refers to the attempt to sequence the 3 billion base pairs that comprise the human genome. In other words, this would be equivalent to being able to describe the genetic essence of a human being using a huge 3 billion figure written with four digits: the four nitrogenated bases. On the other hand, it is estimated that the human genome has roughly 30 000 genes which form a minority part of the total DNA.

It is clear that the sequencing work would be pointless if there were no knowledge of the purpose the deciphered sequences serve. For this reason, structural genomics – sequencing and nothing else – soon gave way to functional genomics, which tries to discover the function of each known sequence. To do this it is very helpful to compare sequences that appear in the genomes of organisms more or less related in evolution (comparative genomics). Comparative genomics helps explain the similarities and differences between the little more than 19 000 genes of the nematode *C. elegans,* the (approximately) 14 200 genes of the fruit fly *D. melanogaster* and human genes. The joint announcement made by Drs Venter and Collins in June 2000 that the human genome had almost completely been sequenced signalled the "end of the beginning". The time of functional genomics or proteomics has come, giving rise to what is already known as the Human Proteome Project, whose name indicates that the goal is to identify the proteins that the sequenced genes code for, and analyse their functions and interactions.

The Human Genome Project opens the doors to a new form of medicine – genomic medicine, which includes pharmacogenomics – which will certainly be of great benefit to humankind. Nevertheless, we are all aware of the important ethical and legal problems that are created, in terms of prediction (predictive medicine), diagnosis, privacy (labour relations and insurance), patenting of human genes, legal identification, and so on.

Bases:
the genome of every living being is composed of a succession of chemical elements or nucleotides, which form the double helix of DNA by bases and which are made up of 3 subunits: a phosphate, a sugar and a base from amongst the following: adenine (A), guanine (G), cytosine (C) and thiamine (T). The two strands of the double helix are always linked by the AT or GC base pairs.

The Human Genome Project[1]

The Human Genome Project (HGP) is an international research effort to chart and characterise the human genome, the entire package of genetic instructions for a human being. This entails laying out – in order – the 3 billion DNA letters (or base pairs) of the full genetic code.

Human genome projects are undertaken concurrently in many countries, including China, France, Germany, Italy, Japan, the Russian Federation, Switzerland, and the United Kingdom, and are co-ordinated with the American effort through the Human Genome Organisation (HUGO), whose members include scientists from throughout the world. The research institutions participating in the project are currently generating high quality, accurate sequencing of the human genetic code for scientists everywhere to use as a no-cost resource without restrictions.

In June 2000, the HGP International Consortium, which includes scientists from the above-mentioned countries, announced that it had assembled a working draft of the sequence of the human genome, which meant that 97% of the human genome had been mapped, and 87% accurately sequenced. In a related announcement the private sector company Celera Genomics, of which Dr Craig Venter is the president and chief scientific officer, also announced that it had completed its own first assembly of the human genome DNA sequence. The results of the public project can be accessed free of charge by academics using the information from Celera Genomics; companies wanting to develop a commercial product from Celera's sequencing database must pay a subscription.

In February 2001, the researchers published their findings in *Nature* and *Science* that the number of genes in the human genome had been established as being between 25 000 and 40 000, and probably in the low 30 000s,[2] fewer than had originally been thought (100 000) and only 4 000 more than the garden weed called thale cress.

The human genome: genetic human nature and diversity

1.
HGP website: http://www.nhgri.nih.gov/HGP/

2.
See *Appendix II* for the websites of *Science* and *Nature*.

The debate on the human genome should, at some point, consider the genetic nature of the human being and human diversity. From a biological and genetic approximation, the processes of hominisation and humanisation are particularly important. The first refers to the concept of the human species and the second to the human being. In other words, what is it that genetically defines the human being? When does individualised human life begin? In the present context, reference will be made only to the first question.

Hominisation: concept and singularity in the human species

According to Unesco's Universal Declaration on the Human Genome and Human Rights of 1997,[1] the human genome underlies the fundamental unit of all members of the human family, as well as the recognition of their inherent human dignity and diversity. The human genome, evolutionary by nature, is subject to mutations that are the origins of human diversity.

When we observe the living world that surrounds us, we do not hesitate to distinguish at first glance dogs from cats and from chickens, and so on. Nor do we hesitate to recognise an individual as a human being, in other words, in biological terms, as an individual belonging to humanity. In any of these cases, we are using a simple morphological criterion to classify each one of these individuals into its corresponding species. We have not needed to know how many chromosomes their cells have, or to characterise their proteins or DNA; we have simply implicitly classified them into their respective groups.

The biological concept of species, which began to emerge in the writings of Buffon and other naturalists and taxonomists of the nineteenth century, refers to the fact that species consist of populations. Furthermore, these species have an internal reality and cohesion due to the genetic programme that has evolved historically and which is shared by all members of the species, such that they constitute a reproductive community, an ecological unit and a genetic unit (Mayr 1970). As a reproductive community, the individuals of an animal species interact with each other as potential sexual partners and seek out other individuals for the purpose of reproduction. There are many mechanisms which ensure the intraspecies reproduction of all organisms. As an ecological unit, the species – regardless of the individuals who form it – interact as a unit with other species, with whom they share the same environment. As a genetic unit, the species is formed of a large intercommunicating genetic pool for a short period of time.

1.
The text of Unesco's Declaration can be accessed at: http://www.unesco.org/human_rights/hrbc.htm

35

These three cited properties – according to Mayr – elevate the species above the typological interpretation of a simple "class of objects" and lead to a biological concept of species as "groups of interbreeding natural populations that are reproductively isolated from other such groups". In short, a species is a protected gene pool; it is a Mendelian "population" (in the broad sense of a set of populations) that has its own means (isolation mechanisms) to protect itself from the harmful flow of genes from other gene pools. Genes from the same gene pool form harmonious combinations arising from a co-adaptation produced by natural selection during the evolutionary process. The mixture of genes of different species leads to a high frequency of non-harmonious genetic combinations with the result that the isolation mechanisms that impede such a mixing are favoured by natural selection.

Even accepting, with its limitations, this biological concept of a species, the problem we are dealing with remains, because what really interests us is knowing what it is that characterises the individuals of the human species and what differentiates them from other species. There can be no genetic flow between the human species and that of the chimpanzee, the pongid nearest to us in evolutionary terms, because there is a reproductive isolation between the two that does not give us information on the essential difference between a human being and a chimpanzee.

In the evolutionary process, intelligence appeared thanks to an uninterrupted genetically determined series of anatomical changes that favoured the progressive development of the brain (cerebralisation), so that from a certain moment on, the brain of the hominid was able to carry out intellectual activity, and could apprehend the medium not just as a simple stimulus but also as a reality product of a reflection. When this capacity for reflection was directed towards the individual himself he became conscious; the hominid had reached the critical point in his hominisation, reaching the category of man.

When did hominisation occur? Or, in other words, when was the transition from simple animal who "thinks" to rationality? From the Australopithecines, passing through *Homo habilis,*

Homo ergaster, Homo erectus, Homo antecessor until reaching modern *Homo sapiens,* genetic changes have been accumulating the genetic information that differentiates the individuals of the human species from any other animal species: human beings are genetically capable of being a cultured subject, an ethical subject and a religious subject.

A human being is genetically capable of using symbolic language, understanding by symbol any object or act whose meaning is not evident on its own but rather is socially recognised. Culture arose as a consequence of symbolic language and the evolution of mankind, resulting from the biological evolution and the cultural evolution, started at the very moment in which the first hominid was able to use symbols as a means of expressing his ideas. Given that symbolic language is exclusive to the human species, culture is the domain of humans: a human being is a cultured subject.

The human being is the only creature able to anticipate long-term events and consciously make a choice of action in consequence; in other words, they are genetically capable of making value judgments and of distinguishing good from evil, opting for one or the other (ethical subject: Waddington's "ethical animal"). Waddington (1960) considered that humans are genetically determining "ethicising beings" and, especially in childhood and youth, "authority acceptors".

As a consequence of the evolutionary process, a human being attains the status of a being conscious of himself/herself, conscious of death. Burials and funeral rites appear to be specifically human. Other animal species dispose of their dead kin in the same way that rubbish from the nest or from the den is disposed of, or they just eat them. The fact that animals can sense death does not mean to say that, being healthy and in full youth, they are conscious that they are going to die sooner or later.

This self-consciousness and the consciousness of death no doubt leads human beings to wonder about their reasons for being, about the meaning of life: where they came from and where they are going; this is the basis of religion. Biologically speaking, it is clear that mankind has not inherited any

Monozygotic: derived from a single ovum.

Mutation: the process by which a gene or some other DNA sequence undergoes a change in structure.

Polymorphism: difference in DNA sequence among individuals.

Nucleotide: fundamental chemical building block of DNA.

religion, but has inherited the genetic capacity to be a religious subject and to search for an answer to the mystery; the transcendence, the need to relate to a superior being: God. Human beings are genetically capable of accepting or rejecting religious considerations.

The human genome and diversity

The human genome is always referred to in the singular, as if it were a unique reality. However, we know that, with the exception of monozygotic* twins, the genomes of two individuals cannot be considered as identical. Should we then talk of "human genomes" without being able to generalise the concept? In this context it would be helpful to point out that, in generic terms, when we speak of the human genome we understand the set of DNA present in a complete chromosomal set that is common to all members of the human species, regardless of the variations that might arise in the genetic information contained at different places. We should not forget the evolutionary meaning that delimits the concept of the genome.

Variability is a consequence of an intrinsic genetic property of the hereditary material: mutations*. Therefore, diversity is inseparable from the concept of the human genome itself, as was indicated in the previous paragraph. For this reason, Unesco's Universal Declaration on the Human Genome and Human Rights (Article 3) states: "The human genome, which by its nature evolves, is subject to mutations".[1] In any case, according to the preamble in the United Nations Convention on Biological Diversity of 5 June 1992,[2] it is stated that the recognition of the genetic diversity of humanity should not give rise to any interpretation of social or political order that questions the "inherent dignity and [...] the equal and inalienable rights of all members of the human family".

1. See: http://www.unesco.org/human_rights/hrbc.htm

2. See: http://www.unep.ch/bio/bio-intr.html

Analysis of the human genome has shown the existence of a polymorphism* based on subtle differences between individuals consisting in changes of a single nucleotide*: the SNPs (single nucleotide polymorphisms). The multiple variations of a single nucleotide at different loci may explain both the genetic differences between humans as individuals and between individuals of different populations. Knowledge of the presence of SNPs will be of great help to pharmacogenomics in understanding why the same drug may, or may not be effective, or has, or does not have, side effects when given to different people or why environmental circumstances affect individuals differently in terms of suffering or not suffering from certain diseases. Thus, the Human Genome Project aims to complete a generic or consensus sequencing of the human genome.

Independently, the Human Genome Diversity Project (HGDP) is also being carried out which, in the words of its principal proponent, Cavalli-Sforza (1995), is "an international anthropological project that is trying to study the genetic richness of the entire human species. The project will enhance our knowledge of this genetic richness and will show both human diversity and its profound underlying unity".

The HGDP started its journey in 1991 in a letter which appeared in the journal *Genomics* signed by Luigi Luca Cavalli-Sforza, Allan Wilson (who died shortly afterwards), Mary-Clair King, Charles Cantor and Robert Cook-Deagan. The article defended the need to carry out a systematic study of the human species at a genetic level. In 1994, the Human Genome Organisation (HUGO) decided to adopt the HGDP as one of its research projects on the human genome. The HGDP was financed initially by the US National Science Foundation, the US National Institutes of Health (NIH) and the US Department of Energy.

The aim of the Human Genome Diversity Project is to evaluate the size of the normal human genetic diversity in the human gene pool. Obviously, it is crucial that population sampling is representative of such variation.

It is thought that the human species, from the anthropological viewpoint of diversity, consists of 5000 different populations which correspond to the approximately 5000 languages spoken today which have been classified in seventeen families or phyla (Ruhlen 1987). It is important to emphasise that languages constitute barriers which contribute to the isolation of human populations in evolution. The evolution of mankind results from genetic evolution and cultural evolution.

Marker:
a specific molecule of known properties (eg, size, sequence, etc.) that is associated with a specific disease or characteristic. This is then used as a reference point, against which we can compare unknown samples, to screen for the disease or characteristic.

Cavalli-Sforza et al. (1988) analysed human evolution bringing together genetic, archaeological and linguistic data. Linguistic families correspond to groups of populations with very few overlaps, and their origin can be given a time frame. They found, on the one hand, that average genetic distances between the most important clusters are proportional to archaeological separation times. On the other hand, they also found a considerable parallelism between genetic and linguistic evolution.

The HGDP hopes to take samples from 500 such populations: on the one hand, blood samples from individuals to establish a cellular reservoir for long-term storage which would provide the DNA necessary for detailed genetic analysis; on the other hand, samples taken from 100 to 200 individuals to carry out short-term genetic studies, but which require a larger sample size for statistical analysis.

As Calderón points out (1996), a source of uncertainty in the HGDP is the size and distribution of the sample programmed in order to be representative of the variation in the human species. The sampling criteria proposed by Cavalli-Sforza have some disadvantages. The proponents of the HGDP indicated that the populations to be selected would be those considered as being of special interest for anthropologists. They also wanted to include in the study all ethnic groups that expressed an interest in the project, giving higher priority to those groups that had been studied in greater detail from the anthropological point of view.

In view of this approach, certain questions can be raised: is it known how many basic human groups exist in the world or, in other words, groups that have a significant fraction of genetic variation in their genome that characterises the human species? What degree of ethnicity must a population have in order to be classified as an anthropological population of interest?

In addition to the aforementioned problems which refer to population sampling, we should also mention problems related to "which genetic markers* ought to be used in terms of how representative and how reliable they are?", and "what is the minimum number of these markers required to be able to reconstruct the biological history of humanity?" A broad consensus on this topic would be required.

Historically, most attempts to classify human populations have been based on biological considerations, often as a reflection of a cultural desire for self-affirmation and self-identification. From a genetic viewpoint two types of characteristics can be used in the study of human populations:

1. quantitative morphological characteristics, which have a polygenic hereditary base* and which may be subject to a strong environmental influence in their phenotypic expression*;

2. qualitative characteristics of a molecular nature (proteins or DNA) and simple Mendelian inheritance, such as blood groups, polymorphic protein systems, and the HLA incompatibility system.*

In classical anthropological studies, racial classification was based on morphological characteristics such as skin pigmentation, size and form of the body, certain facial characteristics (for example, cheekbone, nose, eye socket), cranial measurements, and so on. However, as Cavalli-Sforza (1995) points out, visible morphological characteristics that help us to distinguish populations or individuals from different continents are really superficial differences from the genetic point of view. On the contrary, genetic studies of human populations indicate that, on the one hand, the patterns of genetic diversity in human populations are broad and the distribution is ubiquitous and, on the other hand, the genetic differences between individuals of the same population are greater than those that exist between populations.

The HGDP aroused suspicion from a number of different areas, both from the scientific point of view (for example, problems of whether the sampling is representative) and from the ethical point of view (Fleming 1996). It may be significant that Cavalli-Sforza (1995) presented his project in 1994 to Unesco's International Commission for Bioethics and failed to obtain the organisation's backing. For their part, the indigenous peoples of the western hemisphere of the American continent (North,

Polygene:
any of a class of genes that combine to control a quantitative character such as height or skin colour in humans.

Phenotype:
the physical properties produced by gene expression and environmental influences.

HLA (human leucocyte antigen):
any of a complex of genetically determined antigens, occurring on the surface of almost every human cell, by which one person's cells can be distinguished from another's and the histocompatibility (compatibility between the tissues) and genetic likeness of any two people can be established (see page 57).

Genotype:
the particular type
and arrangement of
genes in each organism.

South and Central) expressed their rejection of the HGDP in a declaration made public in Phoenix (Arizona, USA) on 19 February 1995. The reaction by indigenous communities against the HGDP may be due to the fact that, as denounced by Unesco's Bioethics and Population Genetics Subcommittee (1995), the peoples of the Third World in general, and indigenous peoples in particular, have been the subject of investigation by scientists from developed countries. The history of anthropology is based on the study of "exotic races", and the anthropologists acted in accordance with the prejudices of their time. An example of this is the investigation of skull size carried out in the second half of the nineteenth century, in which scientists measured skulls and classified races according to the measurements. In this classification, white people were supposed to be superior to black people.

The defenders of the HGDP say that the project would contribute to a greater knowledge of the nature of the differences between individuals and between human populations and, as a result, "it will help to fight the popular fear and ignorance of human genetics and will make a significant contribution to the elimination of racism".

As Calderón put it (1996), modern anthropological genetics has reached the well-established conclusion that the greatest human genetic variation is polymorphic, not polytypic. It is precisely the discovery of the existence of small genetic differences between so-called "human races" in comparison with the greater intrapopulation diversity that constitutes an argument against the pseudo-science of racism.

Racism implies the superiority of a race (or population) over others, as well as the idea that the mixture of races is prejudicial to the race that claims to be superior. It should be pointed out that racism arose during the nineteenth century, before the development of genetics. The concept of race is based on morphological characteristics of the phenotype that are neither homogeneous nor easily assigned to specific genotypes.* There is no scientific basis for believing that certain races are genetically superior to others.

Racism was born of a pseudo-scientific philosophy that confuses cultural characteristics with genetic ones. The HGDP should help to illustrate the fact that the human genome is the fundamental unit of all members of the human family, as stated in Unesco's Universal Declaration.

Unesco's Universal Declaration on the Human Genome and Human Rights

In the present context, reference will be made only to the genetic aspects of the Universal Declaration on the Human Genome and Human Rights which Unesco's 186 member states approved at the 29th General Conference on 11 November, 1997.[1] This declaration was later adopted by the United Nations in 1998, coinciding with the 50th anniversary of the 1948 Universal Declaration of Human Rights.

The first item of information to bear in mind when evaluating a universal declaration such as the one that concerns us now is that it should fit into the framework of some general principles that may be taken up by countries with very different cultures, philosophies and interests, thus avoiding unnecessary confrontations. For example, even the title of the declaration, which in all the drafts included the terms "...and the rights of the human being", was finally changed to "... and human rights", no doubt to avoid any controversy over the status of the human embryo and when human development may refer to a person.

The first basic principle that the declaration sets forth is that "the human genome underlies the fundamental unit of all members of the human family, as well as the recognition of their inherent dignity and diversity", adding that, "in a symbolic sense, it is the heritage of humanity" (Article 1).

What does it mean when it says, in a symbolic sense, that "the human genome is the heritage of humanity"? As Gros Espiell (1995) points out, the concept of the "heritage of humanity" has a legal meaning that has been incubating slowly for more than a century:

1.
See *Appendix II* for website.

43

cDNA:
a strand of DNA that is an exact sequence match to a given messenger RNA, which codes for a unique protein.

mRNA (Messenger RNA):
an RNA (ribonucleic acid) that carries the genetic code for a particular protein from the DNA in the cell's nucleus to a ribosome in the cytoplasm and acts as a template, or pattern, for the formation of that protein.

the fact that the human genome is proclaimed as the common heritage of humanity reaffirms the rights and duties of each human being over his or her genetic heritage. These rights and duties can neither be transferred nor renounced. It interests the whole of humanity which in turn, being subject to Law and with respect to the legally-organised international community, protects it, guarantees and ensures that it cannot be subject to any appropriation or disposition by any other individual or by any other collective group, whether they be a State, Nation or People.

In Article 2, the declaration includes three main principles:

1. everyone has their own genetic identity in the material form of their genome;

2. one should not enter into genetic reductionism as individuals are more than their genetic characteristics;

3. there can be no discrimination because of the genetic characteristics of an individual.

In Article 3, reference is made to two important genetic aspects: first, the evolutionary and dynamic concept of the human genome, which is constantly changing genetically through mutations; and second, the concept of the development of the individual as a result of the expression of their genotype (genome) in a certain environment and sociocultural conditions.

One of the most heated controversies at present centres on the patenting of human genes, which will be dealt with in Professor Bartha Maria Knoppers' chapter. My only comment on this is that when the declaration was being drafted, there were apparently countries which threatened to vote against it if such patents were condemned in the draft. In the face of these pressures from both sides, the declaration included a fourth article, somewhat ambiguous in my judgment (although it possibly had to be so) which says that "the human genome *in its natural state* (my italics) shall not give rise to financial gains". The question that is posed once more is whether the sequences of human DNA that are to be patented on many occasions correspond to the natural state in which the human genome is to be found; for example, as might occur on trying to patent sequences of copy DNA (cDNA)* obtained from a messenger RNA* present in cells (in other words, corresponding to exons* of expressed genes).* It will be recalled that the Council of Europe's Convention on Human Rights and Biomedicine,[1] opened for signature in Oviedo, Spain, on 4 April 1997, did not refer to the patenting of human genes, passing the buck to the European Union Directive on the legal protection of biotechnological inventions, which was approved in July 1998.[2]

After introducing a number of subtle distinctions, Directive 98/44/EC approved gene patenting. In fact, it includes some articles on human genes, as follows:

Article 3

2. Biological material which is isolated from its natural environment or produced by means of a technical process may be the subject of an invention even if it previously occurred in nature.

taking into account that "biological material" is defined in Article 2.1(a) as "any material containing genetic information and capable of reproducing itself or being reproduced in a biological system".

Even though Article 5, paragraph 1 states that:

[...] the simple discovery of [...] the sequence or partial sequence of a [human] gene, cannot constitute patentable inventions

paragraphs 2 and 3 of the same article state that:

2. An element isolated from the human body or otherwise produced by means of a technical process, including the sequence or partial sequence of a gene, may constitute a patentable invention, even if the structure of that element is identical to that of a natural element.

3. The industrial application of a sequence or a partial sequence of a [human] gene must be disclosed in the patent application.

Finally, Article 9 of the Directive states that:

The protection conferred by a patent on a product containing or consisting of genetic information shall extend to all material, save as provided in Article 5(1), in which the product is incorporated and in which the genetic information is contained and performs its function.

Exon:
any portion of an interrupted gene that is represented in the RNA product and is translated into protein.

Gene expression:
the process by which the information in a gene is used to create proteins.

1.
The full text of the Convention for the Protection of Human Rights and Dignity of the Human Being (4 April 1997) with regard to the Application of Biology and Medicine is available at: http://book.coe.fr/ conv/en/ui/frm/f164 -e.htm

2.
European Union *Directive 98/44/EC of the European Parliament and of the Council of 6 July 1998 on the legal protection of biotechnological interventions*, L 213 Official Journal, (30 July 1998), pp. 0013-0021, can be accessed online at: http://europa.eu.int/ eur-lex/en/lif/dat/ 1998/en_398L0044. html

It is important to emphasise that the directive is not a bioethical declaration but a legal document.

As stated in its preamble, the Unesco declaration is a consequence of the recognition that the investigations into the human genome and its applications have opened up substantial prospects for improving the health of the individual and the health of humanity as a whole. It is also acknowledged that these investigations should respect the dignity and rights of the person, condemning all forms of discrimination based on the genetic characteristics of the individual. For this reason, in Articles 4 and 5 of the declaration, the importance of investigations into the human genome and their application to the health and well-being of humans is addressed, respecting the dignity and rights of the individual as well as equal access to the benefits of scientific progress in biology in general, and genetics in particular.

Some of the principles included in the Unesco declaration refer to the various important aspects of bioethics. Thus, the following needs are identified:

1. to promote, at various levels as appropriate, the establishment of independent, multidisciplinary and pluralist ethics committees to assess the ethical, legal and social issues raised by research on the human genome and its applications (Article 16);

2. to promote education in bioethics, at all levels (Article 20);

3. to make individuals and society conscious of their responsibility in the defence of human dignity in topics related to biology, genetics and medicine (Article 21);

4. to facilitate on this subject an open international discussion, ensuring the free expression of various sociocultural, religious and philosophical opinions (Article 21).

Unesco's Universal Declaration on the Human Genome and Human Rights and the Council of Europe's Convention on Human Rights and Biomedicine constitute a clear demonstration of the extent to which the world is preoccupied with the power of using genetic techniques in biomedical investigation and its applications.

Manipulation of the human genome, bioethics and society

To gain an insight into human genetic manipulation, the items can be systematised according to the different biological organisation levels (molecular, cellular, individual, population) or to the different developmental stages (gamete, embryo, foetus, individual born) at which the genetic manipulations are carried out or when their effects are shown:

Somatic cell: cells of the body of an organism (with the exception of the gametes or sex cells).

Germ cells: ova and spermatozoa (sex cells).

Manipulation of human DNA

Molecular analysis of human genome

☐ Sequencing (Human Genome Project)

☐ Molecular pre-implantational or pre-natal diagnosis

☐ DNA fingerprinting identification (forensic genetics)

Use of human genes

☐ Introduction into non-human organisms

- Production of therapeutic human proteins (bacteria, transgenic animals)
- Direct effects on animals

☐ Human gene therapy

- Somatic*
- Germ*

Manipulation of human cells

- Somatic cells: cell cultures
- Germinal cells
- Interspecific cellular hybridisation

☐ Somatic cell fusion: gene mapping

☐ Interspecific *in vitro* fertilisation: the hamster test

Reproduction and manipulation of human embryos

- Artificial insemination
- Gamete intratubaric: fallopian transfer (GIFT)
- Gamete freezing (sperm, ovocytes)
- The beginning of human life: the status of the embryo
- *In vitro* fertilisation (IVF)

☐ IVFET (conventional, ICSI, spermatids)

☐ Embryo freezing

☐ Embryo selection (pre-implantational diagnosis, sex selection)

☐ Embryo research

☐ Cloning

- Reproductive

- Therapeutic non-reproductive: tissue cultures

Manipulation of human individuals

- **Positive eugenics**

  ☐ Gene transfer: somatic and germinal gene therapy with human or non-human genes

  ☐ Genetic mosaics

  - Human organ transplants

    – Somatic

    – Gonads

- Somatic non-human organ transplants: xenotransplants

- **Negative eugenics**

  ☐ Preventing deficient genetic offspring

  - Genetic counselling

    – Preventing marriage of couples with genetic risks

    – Control of birth rate

      – Preventing pregnancy (contraceptives, IUDs, the morning-after pill)

      – Sterilisation (male, female)

  ☐ Eliminating deficient genetic offspring

  - Pre-natal diagnosis: eugenics abortion (amniocentesis, chorion villus biopsy, echography, foetoscopy)

  - Infanticide

Manipulation of human populations

- Euphenics

- Man influencing his/her own evolution

As can be seen, the casuistry in human genetic manipulation is very large. Nevertheless, in the present context reference will be made to the Human Genome Project only as a component of human DNA manipulation.

The Human Genome Project and the interdisciplinary social debate

The genetic information that will become known through the HGP will be an important base for medicine in the future; there will be a type of medicine which some have called "predictive medicine" and some "genomic medicine". The final product of the project will be a reference and sequences map that will constitute the "reference book" for human biology and medicine in the coming centuries. Through all of this we should bear in mind that despite the genetic variability of human populations, it appears clear that the quantity and quality of variation does not cause confusion when putting together data and constructing the reference map of the human genome. As Watson (1990) said:

> A more important set of instruction books will never be found by human beings. When finally interpreted, the genetic messages encoded within our DNA molecules will provide the ultimate answers to the chemical underpinnings of human existence. They will not only help us understand how we function as healthy human beings, but will also explain, at the chemical level, the role of the genetic factors in a multitude of diseases, such as cancer, Alzheimer's disease, and schizophrenia, that diminish the individual lives of so many millions of people.

The HGP is going to have important consequences for medical genetics and specifically for genetic counselling. As Jonsen pointed out (1991), the implications of the Human Genome Project on the doctor-patient relationship will produce drastic innovations in diagnosis, prognosis and clinical therapies. The introduction of genetic information on the risks (probability) of some diseases and the inevitability (certainty) of others, as well as the changing severity of these conditions and the multiplicity of factors that influence their expression may alter the terms of the doctor-patient relationship in different ways.

Molecules of re-combinant DNA: hybrid DNA sequences assembled *in vitro* from different sources.

Jonsen referred to these as "the patient as population" and "the in-patient problem".

The first aspect refers to the fact that the patient's family changes from being a community of support to a community of affected persons: families will be considered as focal points for the disease, as carriers of known risks or of inevitable diseases. This will mean that reproductive decisions (whether or not to have offspring) will become as significant in therapeutic thought as the use of drugs and surgery is now. The patient will perhaps be seen less as an individual and more as a "unit of population".

The second aspect – the "in-patient problem" – refers to those cases in which genetic information obtained from molecular analysis allows the person affected to know his or her future long before symptoms of the disease appear, a disease for which effective treatment might not be available at the time. In other words, it will be necessary to distinguish between those patients who do not enter into the clinical world because they are healthy and those in-patients who, although able to enter into the clinical world, cannot be cured.

Faced with the possibility of knowing that a person carries a gene that, over the years, will lead to the development of a disease, this poses the question of who, how, when and to whom should that information be given. On the other hand, what will the psychological reaction of the patient be? One of aggression towards his or her parents? Of personal disillusionment, considering his or her life to be a failure? Of responsibility towards his or her offspring? The knowledge derived from the Human Genome Project may change the saying "a stitch in time saves nine" of preventive medicine to "better to predict than to avoid", which belongs to predictive medicine or genomic medicine.

From the ethical perspective, it is important to remember that with the molecules of recombinant DNA* – the basis of current molecular genetic engineering – an unprecedented situation in the history of science arose: for the first time scientists set a moratorium for their own investigations. A group of pioneering scientists in the new molecular technology, headed by the Nobel prize-winner Paul Berg, and whose members included

other Nobel laureates (Baltimore, Nathans and Watson), published the following manifesto, simultaneously, in three highly
prestigious journals on 19 July 1974, *(Nature, Science* and *Proceedings of the National Academy of Sciences)*:

> The signatories, members of a commission that acts in the
> name and under the patronage of the Assembly of Life Sciences
> of the National Research Council of the United States, propose
> the following recommendations: the first, and most important,
> is that until the potential risk of recombinant DNA molecules
> has been better evaluated, or until suitable methods have been
> developed that prevent their dissemination, scientists through-
> out the world should join this Committee, voluntarily post-
> poning the following types of experiment.

Without doubt, this deed marked an important moment in the
history of bioethics. It can be said that from that time on, in
genetic and biomedical investigation, discussion and interdisciplinary evaluation from the scientific, ethical, legal and social
points of view was occurring at the same time or even before scientific fact became irreversibly established. This is what has
happened with the Human Genome Project: for example, in the
United States it was decided that in the beginning, 3-5% of the
scientific budget would be dedicated to the study of its ethical
and social implications and this attitude has been copied by
other countries (Canada and the European Community decided
to assign 7-8%). As a consequence, an increasing number of
interdisciplinary meetings have been held throughout the
world. In Spain, these have included the meetings held in Valen-
cia in 1990, where ethical aspects were discussed (II Workshop
on International Co-operation for the Human Genome Project:
Ethics), and in Bilbao in 1993, where legal aspects were examined (International Workshop on Human Genome Project: legal
aspects). Both meetings were organised by Professor Santiago
Grisolía, who was a President of Unesco's Human Genome
Project Committee (Fundación BBV Documenta, 1991, 1994).

An international ethical code for human genetics

In 1991, Fletcher pointed out the need to draw up an International Ethics Code for Human Genetics. He based his reasoning
on four premises:

1. The Human Genome Project should and will be successful.

2. Genetic services will gradually come to form part of the national health plans of developed and developing countries.

3. Genetic knowledge will become a normal part of daily life because genetic information will transform medical practice and because the new generation will be educated in the belief that "it is good to want to know" their own genetic characteristics in order to prevent damage to themselves, to their immediate offspring and to future generations.

4. There is already a nucleus of ethics agreements between specialists from many countries on the use of genetic knowledge in the medical and social contexts.

Using these premises as a starting point, Fletcher shows that the ethical problems that genetic clinicians and their patients are facing are not new, but were already present in the time of "old genetics". It is just that now, with the advent of "new genetics", based on domination of nucleic acid technology, the ethical problems have become magnified enormously in scope and complexity.

Fletcher presented the following ethical problems classified in the order of importance below, basing this hierarchy on several criteria: the findings of an international survey, data on frequency of discussion between experts on genetics and on medical ethics, and thirdly, in terms of the number of people whose well-being is negatively affected by the problem:

1. *Equal opportunities for access to genetic services,* but giving preference to those who are at high genetic risk.

2. *Right to abort,* respecting the choice of the parents either to abort or to proceed with pregnancy with an abnormal foetus.

3. *Confidentiality,* although not in absolute terms, when the patient refuses to communicate this to others who might be affected.

4. *Protection of privacy against intrusion from third parties,* establishing effective legal norms that prevent discrimination in the recruitment of workers and the repercussion of

genetic information on life insurance of health insurance premiums.

5. *Dilemmas on revealing genetic data:* to whom and how to report.

6. *Indications for pre-natal diagnosis* only when the health of the foetus is at stake.

7. *Large-scale voluntary or obligatory genetic screening* (of a large percentage of the population), though obligatory only when there is suitable clinical treatment available for the whole population.

8. *Genetic counselling* should always be non-prescriptive.

The undeniable value of scientific advances may be counteracted by the fact that the new knowledge will be a cause of concern only for those affected, unless suitable forms of treatment are developed and these people have access to such treatment. For this reason, Watson (1990) pointed out that "it is imperative that we begin to educate our nation's people on the genetic options that they as individuals may have to choose among". It is clear that this fact, which has implications both for responsible and irresponsible parenthood, may cause additional serious ethical problems.

Bibliography

Bateson, William. 1906. "Report of the Third International Conference on Genetics." Ed. Rev. W. Wilks, *Journal of the Royal Horticultural Society,* London

Calderón, Rosario. 1996. "El Proyecto Genoma Humano sobre diversidad: aspectos éticos." *Rev. Der. Gen. H,* 4:121-39

Cavalli-Sforza, Luigi Luca. 1995. "The Human Genome Diversity Project." Vol. II, IBC Proceedings, Unesco, Paris

Cavalli-Sforza, Luigi Luca; Piazza, A.; Menozzi, P.; Mountain, J. 1988. "Reconstruction of human evolution: bringing together genetics, archaeological, and linguistic data." *Proc. Nat. Acad. Sci.,* 85:6002-06

Fleming, John I., 1996. "La ética y el Proyecto Genoma Humano sobre Diversidad." *Rev. Der. Gen. H.,* 4:159-186

Fletcher, John, 1991. "Etica y genética humana una vez cartografiado el genoma humano." In *Proyecto Genoma Humano: Etica* (co-ord. by S. Grisolía), Fundación BBV, Bilbao, pp. 287-297

Fundación BBV documenta. 1991. *Proyecto Genoma Humano: Etica* (co-ord. S. Grisolía). Fundación BBV, Bilbao, 479 pp.

Fundación BBV documenta. 1994. *El Derecho ante el Proyecto Genoma Humano.* Vol. I, 448 pp.; Vol.II, 287 pp.; Vol.III, 307 pp.; Vol. IV, 399 pp. (co-ord. S. Grisolía). Fundación BBV, Bilbao

Gros Espiell, Héctor. 1995. "El patrimonio común de la Humanidad y el genoma humano." *Rev. Der. Gen. H.,* 3:91-103

International Human Genome Sequencing Consortium 2001. "Initial sequence and analysis of the human genome." *Nature,* 409 (6822) pp. 860-921

Jonsen, Albert R. 1991. "El impacto del cartografiado del genoma humano en la relación paciente-médico." In *Proyecto Genoma Humano: Etica* (co-ord. S. Grisolía), Fundación BBV, Bilbao, pp. 229-39

Lacadena, Juan-Ramón. 1981. *Genética* (3rd edition, prologue). A.G.E.S.A., Madrid

Lacadena, Juan-Ramón. 1992. "Manipulación genética." In *Conceptos fundamentales de ética teológica* (editor M. Vidal), Editorial Trotta S.A., Madrid, pp. 457-92

Lacadena, Juan-Ramón. 1993. "Sobre la naturaleza genética humana." In (F.Abel, C.Cañón, eds.), *La mediación de la Filosofía en la construcción de la Bioética,* Univ. Pontif. Comillas, Madrid, Federación Inter. Univ. Católicas, pp. 15-25

Lacadena, Juan-Ramón. 1996. "El Proyecto Genoma Humano: Ciencia y ética." *Jornadas Iberoamericanas,* June 1996. Real Academia de Farmacia, Madrid, pp. 5-41

Lacadena, Juan-Ramón. 1999. *Genética General. Conceptos fundamentales.* Editorial Síntesis, Madrid

Lander, Eric S.; Weinberg, Robert A. 2000. "Genomics: journey to the center of biology." *Science,* 287:1777-1782

Mayr, Ernst. 1970. *Populations, species and evolution* (Chapter 2). The Bellknap Press of Harvard University Press, Cambridge, Ma.

Ruhlen Merritt. 1987. *A Guide to the World's Languages.* Vol.1. Stanford University Press, Stanford, California

Unesco International Bioethics Committee Population Genetics and Bioethics report. 15 November 1995

Venter, Craig; Adams, M.D.; Myers, E.W. et al. 2001. "The sequence of the human genome." *Science, 291* (5507), pp. 1304-51

Waddington, Conrad Hal 1960. *The ethical animal.* Allen and Unwin, London

Watson, James D. 1990. "The Human Genome Project: past, present, and future," in *Science,* 248: 44-49

Predictive medicine

by Professor Jean Dausset

Aesculapius had three daughters: Hygeia, the goddess of health, Panacea, the goddess of treatment, and Iaso, the goddess of healing. We should perhaps invent a fourth, the goddess of prediction. Whereas, historically, curative medicine preceded preventive medicine and predictive medicine, the order has now been reversed. We must predict in order to prevent and, where necessary, provide early treatment.

One did not have to be an expert on the subject to imagine the prospects opened up in the field of medicine by the discovery that many diseases are associated with human leucocyte antigens (HLA: see below). It was clear that these markers* of susceptibility and resistance would simplify diagnosis and, above all, help in prescribing timely preventive or curative measures.

Marker:
a specific molecule of known properties (eg, size, sequence, etc.) that is associated with a specific disease or characteristic. This is then used as a reference point, against which we can compare unknown samples, to screen for the disease or characteristic.

Alleles:
Variant forms of the same gene. Different alleles produce variations in inherited characteristics, eg, eye colour or blood type.

The Human Leucocyte Antigens (HLA) System

Everyone is familiar with the ABO blood group system. A blood transfusion requires compatibility between donor and recipient. Likewise, under the HLA system, the tissue group must be compatible if organ grafts, and above all bone marrow transplants, are to be successful.

The HLA system is composed of five main genes (A, B, C, DR, DQ) situated next to each other on chromosome 6.

Each of these genes has variants (or alleles*), in some cases large numbers of them (several hundred for gene B).

Every person receives a set of variant genes, one from his father and the other from his mother. It can be seen that the combination of these variants can attain an astronomical figure (several million). Like a bar code, the particular combination is the stamp of individuality of each of us.

The product of these genes is found in the form of HLA molecules on the surface of almost all the body's cells.

Their function is to trigger the body's defences against outside attack by distinguishing between what is alien and what is not.

Lastly, some of these variants are associated with certain diseases, either closely (almost 100%) or less so. With these associations, it is possible to detect those individuals in the population who have a predisposition to contracting a given disease.

This is the basis of predictive medicine.

As early as 1972, I wrote the following: "A relatively important positive or negative association of the HLA determinants with some well-characterised disease would be of the utmost *predictive* value... It is our conviction that tissue alloimmunology will find in this area a new and even larger field of application in addition to its surgical usefulness in transplantation" (Dausset 1972).

Definitions

Medicine can be predictive only if the individual examined does not have any pathological manifestation of a disease for which a possible predisposition is feared.

Thus, we define predictive medicine as "the identification of *healthy* individuals who have a predisposition to develop a given disease. Conversely, predictive medicine also makes it possible to identify individuals who have no such predisposition or are even protected by special resistance genes." This definition excludes foetuses for which a disease has already been detected *in utero* or for which it is known that a disease will develop soon after birth.

The concept therefore concerns all individuals found to be free of the condition feared by a clinical examination or even the most detailed additional examinations.

Hence, the objective of predictive medicine is to identify healthy individuals who are either susceptible or resistant.

Generally speaking, predictive medicine comes before prevention. It identifies, within a siblingship or in the overall population, those individuals who must be monitored preventively. The advantages of this are obvious.

Moreover, the risk for these individuals can even be quantified by comparing in each population studied the number of sick individuals who have one or more genetic risk markers, with the number of individuals who are also sick but do not have some or any of these markers. This is known as relative risk. Predictive medicine is thus primarily probabilistic. But risk is sometimes perceived as a certainty and so can lead to disproportionate fears or even alarm.

Autoimmune disease:
a disease where the body produces antibodies that attack the person's own tissues.

The word "predictive" is sometimes misunderstood, in French at any rate. It conjures up the risks actually incurred depending on the size of the relative risk. Perhaps it would be better to speak of "individual preventive medicine", as opposed to what "until now" has usually been perceived as mass preventive medicine, one example being the universal use of vaccinations.

Diseases linked to the HLA system

Beginning in the years 1970-75, it was discovered that in most areas of medicine, there is an association between many diseases and the HLA tissue typing system (see **Table I**). Sometimes this correlation is spectacular, as in the case of ankylosing spondylitis, a form of chronic inflammation of the spine, which is associated with the B27 variant of the HLA-B gene. Males who have this marker have a 600-times greater risk of developing the disease than those who do not have it. Rheumatologists have used this test for years to help with their diagnosis and treatment, as well as to identify children at risk in affected families.

Two other diseases exhibit a spectacular correlation with the HLA genes. One is the very serious retinopathy (an eye disease) known as "birdshot" (associated with HLA-A29), and the other is the sleeping disorder narcolepsy (associated with HLA-DR2).

Many associations occur in relation to autoimmune diseases* affecting various organs, such as the pancreas (juvenile diabetes), the suprarenals (myasthenia), the nervous system (multiple sclerosis), the skin (pemphigus) or the joints (rheumatoid arthritis). Here the correlation, with variants of the HLA-DR and DQ genes, is often only slight. These genes are not the only

Table I: Diseases associated with or linked to the HLA system

Risk factor higher than 10

Diseases	Associated HLA markers Gene variants	Relative risk
Immuno-haematology – IgA deficiency – alloimmune purpura	B8, DR3 DR3	37 72
Rhumatology – ankylosing spondylitis – Reiter's syndrome – active arthritis	B27 B27 B27	88 37 38
Nephrology – Goodpasture's syndrome – extra-membranous glomerulonephritis	DR2 DR3	16 12
Ophthalmology – anterior uveitis – "birdshot" retinopathy	B27 A29	10 50

Endocrinology
Autoimmunity
– 21-OH deficiency
  – congenital form
  – late form
– insulin-dependent
  diabetes (type I)
– myasthenia
– de Quervain's syndrome
– Gougerot-Sjögren
  syndrome
	B47
	B14
	DR3/DR4
	
	DQA/DQB
	DQA
	B35
	
	DR3
		16
		40
		32
		
		12
		14
		
		10

Digestive pathology – haemochromatosis (HFE) – celiac disease	Mutation C282y/C282y DQA/DQB	344 17
Neurology – narcolepsy – catalepsy	DQB1*0602	483
Dermatology – psoriasis – pemphigus – herpetiform dermatitis	Cw6 DR4 DR3	13 14 15

culprits; instead, the illness is triggered by the unfortunate presence in the same individual of a number of susceptibility genes spread throughout the genome, of which the most important, the *major* one, is in the HLA system.

Polymorphism: difference in DNA sequence among individuals.

In addition to predisposition genes, there are also protector genes, which confer resistance to a particular disease; two typical examples are HLA-DR2 (juvenile diabetes) and HLA-B53 (malaria).

The discovery of some fifty diseases that are correlated with the HLA system, representing only one thousandth of the human genetic heritage, suggested that many other susceptibility or resistance genes ought to be found in the genome's remaining 999 thousandths. To find them, we first had to learn more about the genome. The Centre d'Etude du Polymorphisme Humain (Centre for the Study of Human Polymorphism* – CEPH), which I founded in 1984, carried out pioneering work in this field.[1] Thanks to a generous donation from a major modern art consortium, well before the launch of the Human Genome Project, the centre succeeded in producing the first detailed genetic map and the first physical map of the human genome.

Based on these new data, many teams all around the world set out to identify and isolate the genes responsible for a variety of diseases (whether manifestly hereditary or not), thereby opening the way to possible treatments. But in addition to these specific treatments, an understanding of the genes responsible for predisposition has helped develop a new branch of medicine, predictive medicine, which is no longer confined solely to the HLA genes.

Now that virtually the entire genome has been sequenced, not a week passes in which new pathological genes are not discovered, be they normal genes involved in diseases or mutated genes whose alteration triggers the disease.

Predictive medicine can render great service in detecting the possibility of a late appearance of a disease due to the pathological mutation of a single gene inherited from only one of the parents (so-called dominant monogenic disorders), such as glaucoma or Huntington's disease. In the case of glaucoma, a

1.
http://www.cephb.fr

Recessive disorders:
genetic diseases which develop only if the same genetic defect is passed on by both the mother and the father, as opposed to *dominant genetic disorders* which can be inherited from only one of the parents.

Homozygosity:
having identical alleles at the same point on a pair of chromosomes.

Myocardial infarction:
heart attack.

serious eye disease causing damage to the optic nerve, early medical treatment or surgery is effective, whereas in the case of Huntington's disease, once known as St. Vitus's dance, which leads in adulthood to progressive and fatal dementia, genetic testing makes it possible for individuals to find out that they do not have the defective gene and thus can have offspring without fear of passing on the disease. But until the results are known, close psychological support is needed to prepare for coping with the shock of possibly testing positive.

The case of cystic fibrosis, a common condition that affects the quality of the mucus, primarily in the bronchial tubes and the digestive tract, is particularly enlightening. The gene responsible is very common in European populations, but the carriers remain healthy. To contract the disease, a child must have received the same defective gene from each parent (recessive monogenic disorder).* Fortunately, in France it was recently decided to test all new-born infants systematically in order to start immediate palliative treatment on those affected.

The situation is different for diseases that occur solely in individuals who are so unfortunate as to have a set of genes whose simultaneous presence in their genome creates a susceptibility to certain outside factors in the environment (polygenic and multifactorial diseases). These diseases bring great suffering upon industrialised societies: diabetes (5% of the world population), cardiovascular disease, fat metabolism disorders (cholesterol) and probably many kinds of cancer.

Regarding juvenile diabetes, a very common form of the disease which can have extremely serious consequences, it would be a good idea to proceed as with cystic fibrosis and test all infants for the harmful combination HLA-DR4-DR3 so that those affected can be monitored from birth.

The role in hypertension of two mutations of the angiotensinogen gene (a gene on chromosome 1q42 involved in blood pressure regulation) is well known.

Deletion (loss) transmitted by both parents (homozygosity*) of the gene of the angiotensin conversion enzyme on chromosome 17q23 predisposes to myocardial infarction,* whereas homozygosity of the apolipoprotein* B gene of chromosome 2p24 has been observed to have a protective effect.

An E4 variant or allele of the apolipoprotein E gene results in a much higher predisposition to hyperlipemia (an excess of lipids in the blood) and arteriosclerosis. Its homozygosis leads to an accumulation of "bad" cholesterol and thus to hardening of the arteries. Whereas the homozygous E2 allele affords protection against this condition, a homozygous deletion of the apolipoprotein B gene leads to a high level of cholesterol.

These few predictive associations are merely the vanguard of many others, such as the genes affecting susceptibility to inflammatory bowel disease or iron overload (haemochromatosis) and the many genes affecting susceptibility to the three known forms of cardiomyopathy.

The pharmaceutical industry can be counted on to draw inspiration from this and to mine these new, immeasurable sources of profit, initially by devising the diagnostic tests themselves, and eventually – and, let us hope, as soon as possible for the sake of those affected – by developing a treatment.

This approach may be even more promising in the case of sporadic cancers, whose genetic secrets are beginning to be unveiled. It is known that these forms of cancer are due to an accumulation in certain cells of an alteration in one and then in several chromosomes. Only the hereditary forms of cancer have so far benefited from predictive medicine.

Hereditary thyroid cancer requires ablation of the thyroid for carriers of the harmful gene. But this is not done until tests have been performed to identify the cancerous cells.

The RB1 gene is particularly important in retinoblastoma, a form of eye cancer, which only occurs if both RB genes of chromosome 13q14 are mutated or deleted. Hence the need to detect heterozygotes* in families at risk.

We all know that colon cancer takes a heavy toll, especially among men. For the moment, susceptibility genes have been detected only in forms that run in the family. Colon cancer without polyps accounts for between 4% and 13% of all cases of colon cancer; a gene on chromosome 2 is the culprit. Colon

Apolipoprotein:
a protein that contains and transports lipids in the blood.

Heterozygote:
genomes in which a mutation is present only on one of the chromosomes of the same pair.

Polyp:
a small projecting mass of diseased cells that grows in the body, and is usually harmless.

cancer with polyps* accounts for only 1% of the hereditary forms of colon cancer. The APC gene of chromosome 5q21 is in part responsible. Lastly, familial cancer syndrome is due to a deficiency in a DNA repair gene.

The discovery of the BCRA1 gene on chromosome 17q, responsible for 40% of hereditary forms of early breast cancer and usually accompanied by ovarian cancer, constituted a spectacular breakthrough. Another gene, BCRA2 on chromosome 13, is responsible for other forms of hereditary breast cancer (also 40%) which can also affect men. A protector gene has just been described in the HLA system. The importance of these discoveries is well known: for example, they make it possible to avoid the double mastectomy sometimes recommended for young women.

But the search has only begun: susceptibility genes have just been discovered for prostate cancer and melanoma (skin tumour).

While we may well pride ourselves on the speed at which all these genes have been isolated or cloned, it is surprising that so few preventive treatments are available, since that is after all the aim of predictive medicine.

We can of course step up monitoring, in particular for early signs of cancer, promote health, especially by focusing on diet, or urge people to avoid certain jobs, but we are still only in the early stages of predictive medicine, which in the years ahead should become increasingly effective. However, administrative and psychological precautions must be taken on a large scale to avert certain dangers, because probability-based predictive medicine raises many sensitive issues. It does this at the following three levels:

Predictive medicine at the level of the individual

The individual's autonomy, that is freedom of choice, must be respected. Permission to study a person's DNA must be received, most often in writing. This may be restrictive and include authorisation to conduct only a well-defined test, or it may be broader or even all-encompassing. It must be ensured that the person concerned has as full an understanding as possible of the implications of such a decision. Hence, it is insufficient simply to inform; a non-restrictive, non-directive explanation must be provided that is as clear and complete as possible and is expressed in terms that are consistent with the person's education and degree of understanding. Needless to say, the individual must have the capacity to give legal consent, or else the parents or guardian must do so.

Individual autonomy also means the right to know or not to know the results of genetic tests.

Predictive medicine and the individual within the family and the social environment

In many cases, and particularly in the case of a recessive hereditary disease (transmitted by both parents), a study of the family is needed when the gene, as is still often the case, is not cloned, that is isolated and of known sequence. It can be identified only indirectly in the individual tested through comparison with healthy or sick members of the same family (indirect testing). This shows how important it is to secure the consent of family members. It is also conceivable that individuals will react differently or that, when the findings are made known, personal or family upheavals may occur (divorce, inheritance issues, etc.). Hence the person concerned must exercise great care in speaking with family members and inform them of their right not to know and of the rule of absolute confidentiality. In such circumstances, it is vital for the geneticist to exercise great tact.

Where the harmful genes have been cloned or are in the process of being cloned, study of the family is not essential because the gene can be identified directly in the affected individual. Such testing is direct.

The precautions *vis-à-vis* society as a whole are no less important. Confidentiality must be strict, obviously with the exception of the family physician, who is bound by professional secrecy, which in this case is shared with the geneticist. Likewise, these results cannot be used for scientific research without the express consent of the person concerned and, of course, only under a code number which does not divulge the name of the DNA donor.

There has been considerable discussion on the harmful use that an employer might make of this information to redeploy or even dismiss an employee. On the other hand, this same information might help the employer find a post that is compatible with an employee's susceptibility to a given toxic substance (benzene, asbestos etc.).

But above all, society must be able to take precautions to prevent a fatal accident, for example where there is evidence to suggest that an airline pilot might suffer a heart attack. It would seem that in such a circumstance the lifting of professional secrecy would be warranted.

Another widely debated subject is that of insurance. Needless to say, there is a risk that insurance premiums will be raised in response to unfavourable genetic tests. It might therefore be argued that the results of such tests should not be communicated to the insurer, even when requested. But we see no obvious difference between so-called "biological information" requested by an insurer and genetic tests. Moreover, when setting premiums, insurers do take into account such notions as obesity, which is very often of genetic origin. There is no easy way out of this quandary, unless it is to adopt a drastic solution, namely to base insurance premiums on age alone. That is perhaps totally unrealistic, but it would be an act of solidarity by the healthy towards the less healthy members of society.

Predictive medicine at the level of all of humanity

It is argued that predictive medicine might eventually endanger the genetic balance of a population. Admittedly, it will enable more healthy heterozygotes (carriers of a defective gene) to have children. This will increase the frequency of the gene in the overall population, and hence the risk of marriage between two heterozygous spouses with the same defective gene, a quarter of whose children would then be homozygous, that is sick, and would be a burden on society. But this new "genetic burden" would probably be relatively slight.

As predictive medicine will most often concern polygenic and multifactorial diseases, it will usually target adults who in most cases have already had children, and so the impact on the population's gene pool will be very slight. Let us not forget that a gene as such is neither good nor bad.

These are the early stages of predictive medicine, and it is obviously still faced with many obstacles and limitations. It is based merely on probability, which is often difficult to quantify, especially in polygenic diseases, and even in monogenic diseases of variable penetrance. Moreover, a preventive or therapeutic treatment does not always exist. If one does, it may be only partly effective or even have undesirable side-effects when an apparently healthy individual is being treated. Lastly, the cost of the test might be prohibitive at the current time.

The French Academy of Medicine recently made an important recommendation aimed at preventing the premature or improper use of predictive tests: commissions should be set up with geneticists and specialists from all the major medical disciplines (for example, cardiovascular, neurological and oncological diseases) to decide on whether and how to implement a predictive test for a given disease and to discuss the ethical issues raised.

Despite these obstacles and limitations, many of which are temporary, predictive medicine will gradually spread to all areas of pathology thanks to spectacular progress in human genetics.

What impact will predictive medicine have on the everyday practice of medicine in the future?

Predictive medicine should eventually change the nature of medical consultation. Physicians in the twenty-first century will gradually become counsellors. They will help their healthy "patients" to remain so and to manage their health capital in the long term, just as one advises on the management of a portfolio. This careful monitoring should make it possible to realise one of humanity's old dreams, of achieving – by the year 2050, in the richest countries at least – a situation that will reflect a long life with no physical or mental distress and which ends gently in natural death at 100 or 120 years of age. In closing, let us cite the beautiful words of Maurice Tubiana on predictive medicine: "[It] gives physicians a new obligation: to educate those who are well . . . Health is not a gift, it must be conquered; ultimately, it is a duty to oneself . . . and to others."

Bibliography

Dausset J., "Correlation between histocompatibility antigenes and susceptibility to illness", in *Progress of clinical immunology*, R. S. Schwartz, Ed. Grune & Stratton, 1972, I, 183-210

Feder J.N., Gnirke A., Thomas W. et al. "A novel MHC class I-like gene is mutated in patients with hereditary haemochromatosis." *Nature Genetics,* 1996, 13, 399-408

Fernandez-Arquero M., Figueredo M.A., Maluenda C., De la Concha E.G. "HLA-linked genes acting as additive susceptibility factors in celiac disease." *Hum. Immunol.,* 1995, 42, 295-300

Gerbase-Delima M., Pinto L.C., Grumach A., Carneiro-Sampaio M.M.S. "HLA antigens and haplotypes in IgA-deficient Brazilian paediatric patients.» *Eur.J. Immunogenet.,* 1998, 281-285

Khalil I., Deschamps I., Lepage V., Al-Daccak R., Degos L., Hors J. "Dose effect of Cis-and trans-encoded HLA-DQαβ heterodimers in IDDM susceptibility.» *Diabetes,* 1992, 41, 378-384

Khalil I., Berrih-Aknin S., Lepage V., Loste M.N., Gajdos P., Hors J., Charron D., Degos L. "Trans-encoded DQαβ heterodimers confer susceptibility to myasthenia gravis disease." *Life Sciences,* 1993, 316, 652-660

Lechler R., Warreus A. *HLA in health and disease.* Second Edition, Academic Press, 1999

Marcadet A., Gebuher L., Betuel H., Seignalet J., Freidel A.C., Confavreux C., Billiard M., Dausset J., Cohen. "DNA polymorphism related to HLA-DR2 Dw2 in patients with narcolepsy." *Immunogenetics,* 1985, 22, 679-683

Merryweather-Clarke A.T., Pointon J.J., Shearman J.D., Robson K.J.H. "Global prevalence of putative haemochromatosis mutations." *J. Med. Genet.,* 1997, 34, 275-278

Gene therapy

by Professor Robert Manaranche

The sequencing of the human genome, soon to be totally complete, opens up several possible new approaches of which gene therapy is the best known. The genomes of numerous pathogenic organisms (able to cause disease), already sequenced or currently being sequenced, also open up many new prospects for combating infectious diseases. Here, we are concerned only with the implications of knowledge of the human genome, first for the treatment of genetic diseases, and second for the treatment of acquired diseases.

Prion: an infectious particle that does not contain DNA or RNA, but consists only of a hydrophobic protein; believed to be the tiniest infectious particle.

Drug therapy

The information obtained from the sequencing of the human genome provides a great many targets for pharmacology. Recombinant proteins, that is those obtained by genetic engineering, will make it possible to avoid the risks inherent in using proteins of animal or human origin; the prions* associated with BSE and the transmission of Aids through contaminated blood are the most striking illustrations of these dangers. Furthermore, the possibilities offered by modern chemical techniques will undoubtedly lead to the discovery of new molecules which will be active against proteins discovered as a result of progress in genetics and which, when they mutate, are the cause of many illnesses. The chapter on industry and the human genome in this book examines in more detail the contribution by genetics to drug therapy.

Gene therapy, in the broad sense of the term, will be a major part of the contribution made by genetics to treating many illnesses, not only genetic disorders but also acquired diseases such as cancer or serious infectious diseases.

Gene therapy

Modifying the genetic make-up of a human being in order to effect a cure has proved to be much more complicated than

Transgenic animal: animal whose genome has been genetically modified.

was imagined at the outset of this exciting venture (Kahn 2000). Today we have a better idea of the difficulties of the task, and significant progress has been made in overcoming them. There are now options other than simply transferring a healthy gene into diseased cells. Broadly speaking, what we are seeing is the emergence of a new form of medicine which is able to improve the situation of, and in some cases cure, people suffering from serious – and not just hereditary – illnesses.

The concept of gene therapy and its applications

Experiments have shown that it is possible to transfer foreign genetic material into a cell; this was first demonstrated some fifty years ago with bacteria. And indeed, when a spermatozoon, during fertilisation, deposits its DNA into the ovum, it carries out the most complex transfer of genes imaginable.

The first experiments aimed at eliminating a gene in a mammal (a mouse) were carried out in the early 1980s. Today, modifying the genome of the developing egg of a laboratory mouse is a relatively commonplace laboratory practice. We know how to eliminate a gene in a mouse, thereby creating a "knock-out mouse" and, conversely, how to introduce a new gene and create a "knock-in mouse". These transgenic animals* are extremely valuable for studies on a given genetic disease and for devising therapeutic strategies.

The first attempt to modify the human genome for therapeutic purposes dates back to 1988 and Steven Rosenberg's experiment in the United States designed to treat an advanced cancer. In September 1990, the laboratories of F. Anderson and M. Blaese, also in the United States, carried out the first experiment using gene therapy to treat a genetic disease, on children suffering from a serious immune deficiency. It is still difficult to assess the results of each of these two experiments. In the first case, as indeed in all those which have followed in cancer research, it is impossible with patients in the very advanced stages of the disease to say with any degree of certainty what difference the treatment made and what would have happened if no treatment had been administered. The second experiment was designed to correct a deficiency in the immune system caused by an enzyme (adenosine deaminase – ADA), and the children continued to receive enzyme treatment as a precautionary measure.

With these initial experiments, we can see how attempts at gene therapy have been directed towards two quite distinct categories of disease: cancers and monogenic illnesses (resulting from the mutation of a single gene). In order to halt the tumour process, it is necessary to destroy cells or at least stop them from proliferating. In order to treat a genetic disease it is necessary to restore the correct functioning of diseased cells. The level of difficulty to be overcome is not the same; as in other spheres, it is always much easier to destroy than to construct. It is not surprising, therefore, that the vast majority of clinical gene therapy trials have focused on various forms of cancer. To this must be added the vast experience of cancer researchers in the field of clinical trials and the pharmaceutical industry's interest in disorders affecting a very large number of patients.

Numerous attempts at gene therapy on humans have been made in the last ten or more years. According to the only international statistics published and regularly updated by Wiley & Sons, publishers, there have been, as of 25 May 2000, 425 gene therapy trials on 3476 patients. Of these, 279 (65%) concerned various forms of cancer. Only 55 (13%) concerned patients suffering from monogenic diseases. The remaining 33 concerned patients suffering from an infectious disease, generally Aids.

The main monogenic disorders for which clinical trials have been carried out are cystic fibrosis, severe immune deficiencies linked to the X chromosome (see page 78), haemophilia B (a haemorragic disorder), chronic granulomatous disease (in which an enzyme deficiency prevents the white blood cells from eliminating bacterial and microscopic fungal infections), familial hypercholesterolemia, Duchenne muscular dystrophy (a muscle disease) and Fanconi's anaemia (a serious form of anaemia). Trials have also been carried out on several enzyme deficiency disorders: Hurler and Hunter syndromes characterised by skeletal changes and mental retardation, ornithine-transcarbomyl transferase deficiency, a liver disease causing an accumulation of ammonia in the blood, Canavan disease,

Plasmid:
a small fragment of DNA, usually circular: used in recombinant DNA procedures to transfer genetic material from one cell to another.

Vector:
A molecule (or association of molecules) capable of transporting a substance (eg DNA) to a cell.

involving serious psycho-motor retardation, and Gaucher's disease which combines anaemia with neurological disorders.

The justification for so many trials is uncertain. Clearly, trials which merely repeat under the same conditions something which has proved to be ineffective raise some doubt, at least, as to whether they are necessary. A number of these trials were made in conjunction with companies quoted on the stock exchange which were keen to publish triumphant press releases right from the very beginning. Accordingly, one might legitimately question the expediency of some of these experiments.

Supervision by independent public bodies is absolutely essential if trials on humans are to be carried out with minimum risk. Accidents (including one fatal one) have been reported in the United States, and stricter control of these clinical trials has proved essential. In Europe, the countries where experiments may be carried out have adopted strict control mechanisms, and this is doubtless the reason why no serious accident has been reported during the 97 trials carried out in Europe.

Research over the last ten years has helped circumscribe the problem more clearly, and led in 2000 to the first undeniable success in the field of gene therapy. This took place in Europe, and more specifically in France, in the department run by Professor Fischer at the Necker hospital in Paris. This highly successful experiment is described on page 78 in greater detail.

The problems raised by gene therapy

In order to obtain the long-term expression of a normal protein, without any undesirable side-effects, in a patient suffering from a genetic disorder, a way has to be found of delivering into a large number of specific cells a normal gene which will express the missing protein.

Two solutions have been used to achieve this: injecting into the organism a plasmid*, which is a fragment of naked or purified DNA accompanied only by its regulatory sequences; or using a vector* which may be a virus or a chemical packaging compound for the DNA to be implanted. This is very much a case of using a Trojan horse to achieve a therapeutic solution.

Whatever the solution chosen, it has to be realised that the fragment of foreign DNA introduced into the body has to survive what can only be described as a veritable obstacle course. DNA carries negative charges and will be repelled by the compact structures, generally negatively charged, which fill the extracellular spaces.[1] A further obstacle to be overcome is for the DNA to break through the lipid bi-layer of the plasma membrane; this is often achieved by a process of endocytosis*. Once in the cytosol*, the DNA fragment has to find the route to a nuclear pore*. Then, once in the nucleus, this DNA has to survive the nucleases which could damage it, and lastly it has to be able to express itself, which it will only be able to do in an effective and lasting way if it can incorporate itself into a chromosome of the host cell.

Plasmids

Numerous experiments on animals since 1990, primarily in J.A. Wolff's laboratory at the University of Wisconsin in Madison (USA), have shown that an intramuscular injection of a fragment of circular DNA (a plasmid) could penetrate cells and express the protein for which it carried the code. The problems to be overcome include ensuring that there is broad diffusion beyond the injection point and that there will be lasting gene expression*.

A number of solutions have been put forward to facilitate plasmid diffusion. One is the application of short electrical discharges (pulses) close to the point of injection. This is referred to as electroporation. The advantage of this procedure, above and beyond the fact that it improves diffusion, is that it enables the plasmid to penetrate the cells. Very encouraging results have been obtained using this technique with laboratory animals; not only was diffusion achieved over several centimetres but the "transgenesis" was stable for several months – up to nine months in the most effective experiments.

Another possible means of ensuring diffusion, tried out in *in vivo* experiments on animals, is the use of substances facilitating the passage of plasmids through the extracellular areas which inhibit their progress. Much use has been made of

Endocytosis: the process whereby a cell incorporates foreign material by folding inward a portion of the cell membrane.

Cytosol: the water-soluble components of cell cytoplasm.

Nuclear pore: pore of the membrane of the cell nucleus through which molecules can pass in both directions.

Gene expression: the process by which the information in a gene is used to create proteins.

1.
See page 138.

hyaluronidase (an enzyme extracted from spermatozoa, where it enables the male gamete to penetrate the membrane surrounding the ovum). Proteins which are collagenases (those capable of destroying collagen), such as metallo-proteinases, have led to improved plasmid penetration. These chemical techniques have improved diffusion but have a number of serious disadvantages: the cohesion of healthy cells can be adversely affected in the same way as defective cells. In all, plasmids have been used in 16 clinical trials involving 71 patients.

It is possible to use plasmids intramuscularly, or more generally via other specified tissues, but further improvement is necessary to secure broader diffusion and long-term expression. Preliminary experiments on delivery via the blood, for example via the hepatic vein, to transfect (enable a DNA fragment to penetrate) liver cells have been successfully carried out on animals. However, transposing this to humans raises serious problems of ensuring that the process is completely safe.

Vectors

In order to deliver foreign DNA with a stable expression into an organism to produce a therapeutic effect, in most cases a "package" of this DNA fragment is used in a vector. This may be either a virus or an artificial vector – a chemical compound, generally lipid-based. The problems of vectors are at the very heart of the difficulties to be resolved in order to ensure effective gene therapy. A new science has emerged: vectorology, on which several laboratories throughout the world are working. In France, the Généthon laboratory, set up by the Association Française contre les Myopathies (French Association to combat muscle disorders) with funds from the Téléthon campaign, has set this as its priority objective, and in conjunction with a network of laboratories provides European researchers with vectors which satisfy the quality criteria required for clinical trials, but which are intended simply for pre-clinical experimental work.

Several conditions – constituting a veritable technical specification – have to be met for these vectors to be able to be used in gene therapy. They must be totally innocuous, have sufficient insert size for the transgene (foreign gene incorporated into the cell genome), and not be immunogenic, that is not provoke an immune reaction; they must ensure prolonged transgene expression; and lastly, they must target only the cells to be corrected. Finally, the vectors must be produced in large numbers under the strictest safety conditions essential for all human experiments.

Viruses as vectors

Viruses are natural vectors because, throughout evolution, they have acquired the capacity to introduce themselves into the cells outside which they are unable to survive. Several types of virus have been used in gene therapy.

Retroviruses: These viruses, whose genetic material consists of RNA (ribonucleic acid), have been used primarily in tumour suppression because of their ability to penetrate, generally speaking, only dividing cells. The most commonly used retroviruses to date have been those causing leukaemia in mice (MLV). It is now possible to produce them in large quantities, using techniques which satisfy the quality criteria for clinical trials. By using specific promoter sequences and because of the special characteristics of the envelope, it is possible to obtain tissue specificity permitting a certain degree of targeting.

Their main limitation is the fact that the majority of retroviruses cannot be used for non-dividing cells, or cells which divide only to a very small extent (muscle cells, neurones, hepatocytes etc.).

However, one family of retroviruses – lentiviruses, such as HIV, the virus responsible for Aids – is able to penetrate non-proliferating cells. These lentiviruses give prolonged transgene expression, sometimes beyond 7kb (kilobase = 1 000 bases*), probably by chance integration into the genome. Much work has been done to ensure that the lentiviruses used as vectors in gene therapy are completely harmless. Today, safety levels are such that discussions are currently being held, or are about

Bases:
the genome of every living being is composed of a succession of chemical elements or nucleotides, which form the double helix of DNA by bases and which are made up of 3 subunits: a phosphate, a sugar and a base from amongst the following: adenine (A), guanine (G), cytosine (C) and thiamine (T). The two strands of the double helix are always linked by the AT or GC base pairs. The base pair has become the unit of measurement for every genome. One kilobase (kb) represents 1 000 base pairs. The human genome has over 3 billion base pairs. The genome of a virus has between 1 000 and 100 000 base pairs (or 1-100 kb).

to be held, with the authorising bodies for the first trials to be carried out on humans.

Adenoviruses are viruses responsible for benign respiratory tract infections, particularly rhino-pharyngitis. They are able to penetrate proliferating and non-proliferating cells. They do not insert their genetic material into the chromosomes of the host cell. Their genome comprises a dozen or so genes which can today be neutralised to provide a higher insert size (around 28kb) for foreign genes. First-generation adenoviruses had a limited transgene expression capacity of 3 to 5 months. A much longer period of 10 months or more can be achieved today. However, their use is limited to cases in which a transient expression level is acceptable. One serious disadvantage of human adenoviruses is that a large proportion of the population (90-95%) are immune to this vector. A number of canine-origin adenoviruses have been developed which elicit no immune reaction. Work is currently under way to produce canine viruses with, like human viruses, a genome in which many of the genes have been deleted (gutless viruses) in order to increase their therapeutic DNA insert size.

Adeno-associated viruses (AAV): this term covers a category of vectors comprising a small virus of the parvovirus family dependent on a helper virus, usually adenovirus, to infect cells. These non-pathogenic viruses have only a moderate immunogenic effect. Their DNA is made up of 4 680 nucleotides which, after they have been eliminated (neutralised) can accomodate 4.2kb for the DNA that will transfect the cells to be corrected. The gene introduced by the AAV into a cell integrates preferentially into a specific point on the long arm of chromosome 19, although the mechanism behind this is not yet completely understood. This capacity for integration into the host cell's genome makes it possible to obtain the prolonged expression of a therapeutic gene after a single injection. Adeno-associated viruses have so far been used in 4 clinical trials covering 36 patients.

This vector therefore presents many advantages, but also two drawbacks: the small size of the foreign DNA which it can transport and the difficulties in preparing it in large concentrations.

76

Other viruses have been used in both pre-clinical and clinical trials as vectors to correct a given disease. Two examples at least are worthy of mention: the herpes virus, for which there has been a relatively large number of pre-clinical trials because of its large foreign gene insert accommodation capacity. Pox viruses, large DNA viruses, which are responsible for various illnesses in humans such as smallpox and vaccinia (cowpox), have been used in 25 clinical trials on 130 patients, mainly for different forms of cancer.

Synthetic vectors

Much work in gene therapy has been devoted to the use of vectors made up of various chemical compounds. Lipids have been used in particular because of their ability to produce microspheres, limited by a lipid bi-layer, comparable to that of cell membranes. The positive charges carried by these compounds have high affinity for the negatively charged DNA and at the same time provide a high capacity for cell penetration. These liposomes are capable of packaging large-sized DNA.

Very many formulae have been proposed and tried out, some of which have made it possible for the genetic material to penetrate the cell nucleus. The integration of exogenous DNA in the chromosomes of the host cell remains an unsolved problem, and attempts to overcome it have been made by adding viral proteins and associated enzymes to the synthetic vector.

The advantages of synthetic vectors for gene therapy are significant: there is no limitation on the size of the transgene, it is possible to incorporate on to the vector wall various compounds to target and integrate the genetic material on demand, and they are able to overcome immune reactions. Significant efforts are being made to ensure that perfectly controlled material is available, very close to the requirements for pharmaceutical products. Unfortunately, as yet in the 75 trials carried out, it has not been possible to find an effective system guaranteeing lasting expression of the injected material.

Stem cell: cells which can differentiate into the various types of cell in the adult organism: neurones, muscle cells, liver cells, etc. Stem cells exist not only in the embryo but also in the adult, as recent discoveries have shown.

Gene therapy associated with cell therapy: the first real success of gene therapy

One avenue which seems to be among the most promising is that of gene therapy on stem cells* which, following *ex vivo* culture and transgenesis, are reintroduced into the organism. The only real success obtained so far in gene therapy has been achieved by Professor Alain Fischer's laboratory at the Necker hospital in Paris, using this approach (Cavazzana-Calvo et al. 2000).

The five young children treated were all suffering from a severe immunodeficiency disorder linked to the X chromosome (see below). Cells taken from the children's bone marrow were transfected by a retroviral vector, then cultivated and reinjected five days later into the blood system. The T and NK white blood cells reappeared in the children after a variable period of time. Because of their restored immune defences, the children could be brought out of their protective bubbles after three months; four of them were able to leave hospital a few months after the gene transfer and now live with their families without any treatment. The fifth, more seriously affected, has had to stay in hospital for further treatment.

The researchers have not ruled out the possibility of a recurrence of the problem in the future, but at that point the treatment could be repeated.

Chromosome X-linked disorders

The X chromosome plays a particular role, as there is only one in men (who are XY) whereas it is present in the form of a pair in women (who are XX). A mutation on the single X chromosome in men can be the cause of a genetic disease, whereas with women there is a built-in compensation mechanism: statistically, in half of the cells there is a "healthy" X chromosome. Generally speaking, women are carriers of, but do not suffer from, genetic defects in the X chromosome. Over generations, pairings of two mutated X chromosomes may occur, and women can therefore be afflicted with a disease carried by the X chromosome. Consequently, genetic diseases which are relatively frequent in men are vary rare in women.

Pre-clinical experiments on stem cells

Research on mice carried out primarily in Italy and the United States has shown that bone marrow stem cells were able to regenerate or differentiate into muscle cells after re-injection into the blood (Ferrari et al. 1998; Gussoni et al. 1999). Canadian researchers have also shown that bone marrow stem cells in the rat were able to differentiate into heart cells (Wang et al. 2000). This discovery should make it possible to repair the damage caused by coronary thrombosis.

Stem cells have been found in adult organisms elsewhere than in the bone marrow, especially in the brain. This offers numerous therapeutic possibilities for degenerative diseases, particularly of the nervous and muscular systems, which are major causes of invalidity.

The ideal solution, and much work has been put into this, would be to use gene therapy to correct *ex vivo* a patient's stem cells, which would then be re-injected. There is much work still to be done, but using gene therapy for the prolonged correction of cultured cells is clearly much easier than correcting entire tissues. Autografts should not pose any immunity problem. Among the remaining problems to be resolved are those relating to cellular biology, particularly for cultivating, in sufficient quantity and in complete safety, cells taken from an adult organism.

While gene therapy offers enormous possibilities for treating genetic diseases, a great deal of laboratory work still has to be done to ensure that it can be effectively applied in complete safety.

DNA repair techniques

Gentamicin and stop codons*

It has been known since 1985 that certain antibiotics of the aminoglycoside class affect bacteria by binding with a specific ribosomal RNA site to enable translation to proceed past a stop codon. In 1996 and 1997 researchers came up with the idea of using this characteristic to correct the expression of the cystic

Stop codon: any of the three mRNA sequences which doesn't code for an amino acid and which signals the end of protein synthesis. See *Appendix I:* proteins.

Oligonucleotide:
a chain of a few nucleotides (a nucleotide is an elementary chemical building block in DNA).

Chimeraplast:
a synthetic molecule made up of both RNA and DNA which is used in gene repair.

fibrosis gene in patients having a nonsense mutation (that is one which introduced a stop codon). A mutation of this type is found in only 5% of all patients suffering from cystic fibrosis, but in certain endogamous populations the proportion can be as high as 60%.

Recently, US researchers (Barton-Davis et al. 1999) working with mice as models for Duchenne muscular dystrophy carrying a stop codon were successful in obtaining an expression of dystrophin, the deficient protein in this disorder, at 10-20% of normal levels, using gentamicin, a molecule hitherto used as an antibiotic.

Unfortunately, gentamicin can cause deafness and occasionally renal failure; a non-toxic dosage therefore has to be worked out. Furthermore, there is clearly the risk that it will affect any stop codon. The gene to be corrected therefore has to be targeted; this could doubtless be done by preparing compounds recognising the sequences before and after the stop codon in question.

This technique could have a wide field of application since it is estimated that there are around 600 genetic disorders due to a stop codon.

Chimeraplasty

Chimeraplasty is a technique to correct mutated DNA which first emerged in 1996 from the work of E. Kmiec, who was successful in correcting a specific mutation by transfecting cultured cells using a "chimeric" oligonucleotide*, so called because it is composed of DNA and RNA (Kmiec 1999).

A small fragment of double-stranded DNA, comprising the sequence to be re-established, is combined with a fragment of corresponding RNA in the areas adjacent to the site of the mutation, forming what is known as a chimeraplast.* Promising results were first of all obtained *in vitro* on cell lines to correct a haemoglobin mutation responsible for sickle cell anaemia, a serious blood disorder, and on liver cells in order to correct the gene of the enzyme alkaline phosphatase. Subsequently, experiments were carried out *in vivo* on rats by intravenous injection to correct a mutation of coagulation factor IX, responsible for haemophilia B.

More recently, it has been shown that it is possible to correct neuromuscular disorders, first of all in a murine model and subsequently in a canine model. In the latter case, researchers were able to produce a stable correction over forty-eight weeks. As yet, the results have not yet led to the possibility of physiological improvement, but they have raised well-founded hopes which should be confirmed by improved techniques.

Antisense RNA

An antisense RNA is a complementary RNA* which can bind with a target RNA and thereby inhibit gene expression. This process was initially proposed in 1978 to inhibit Rous sarcoma cancer virus replication inhibitors. Over the last twenty or more years, several applications have been proposed (Dunck-ley et al. 1998). A number of companies market different anti-sense RNAs for therapeutic purposes. In order to be effective, the antisense RNA fragment must satisfy a number of conditions: it must be resistant to nucleases, be capable of crossing the cell membrane and be specific for binding to its target sequence.

This type of treatment has been used to inhibit the expression of proteins which play a part in cancer development. Antisense RNAs have also been suggested as a means of treating psoriasis and even of inhibiting the Aids HIV virus.

Gene immunotherapy

The principle behind this form of immunotherapy is to stimulate the immune system by an intramuscular injection of naked DNA. This type of vaccine has mainly been used to treat tumours by overexpression* of normal proteins or expression of abnormal proteins which will result in an immune reaction against cancer cells (Old 1996). Several products have been marketed, in particular by Vical in the United States under the name Allovectin®, which uses a gene encoding the rare HLA-B7 antigen and which has proved relatively effective against metastatic melanoma, and Leuvectin™, another Vical product,

Complementary RNA: synthetic RNA (ribonucleic acid), whose sequence is complementary to an RNA sequence or a single-stranded DNA sequence.

Overexpressed gene: one which is transcribed in much higher quantities than normal, which usually leads to an increase in the protein that the gene codes for.

which contains the gene encoding the IL-2 interleukin, used to fight metastatic kidney cancer. Unfortunately, these products present a degree of toxicity for certain patients.

There has also been research based on this technique to find vaccines for infectious or parasitic diseases, in particular the HIV virus which can lead to Aids or parasites such as the malaria agent *Plasmodium falciparum*. The US company Copernicus Therapeutics has recently signed a collaboration agreement with the US Naval Medical Research Center (NMRC) to develop a vaccine of this type to combat the latter disease.

The decoding of the genome of various organisms, from viruses to man, provides an enormous amount of information giving a greater insight into biological phenomena and the way they can malfunction with pathological results. Various therapeutic strategies have been devised and in many cases tried out on the basis of this information. Gene therapy in the broad sense is now beginning to emerge successfully from the virtual world. For its application to a large number of diseases, more sustained effort is required, particularly on the development of effective vectors.

DNA repair techniques are in their infancy, but there is hope for real therapeutic applications. A more personalised approach to pharmacological treatments will no doubt offer improved treatment for serious and common diseases.

This rapid overview has shown that there are avenues opening up to improve patients' lives. The United States has often occupied a leading position in the therapeutic field; clearly the countries of Europe, with their considerable research potential and their pharmaceutical industries investing heavily in these new forms of treatment, will play a large part in advancing this new medicine, which promises progress as considerable as that made in the age of Pasteur or following the discovery of antibiotics.

References

Barton-Davis E.R. et al.: "Aminoglycoside antibiotics restore dystrophin function to skeletal muscles of mdx mice", *J. Clin. Invest.* 1999; vol 104: 375-381

Cavazzana-Calvo M. et al.: "Gene therapy of human severe combined immunodeficiency (SCID)-X1 disease", *Science* 2000; vol 288: 669-672

Dunckley M.G. et al. "Modification of splicing in the dystrophin gene in cultured Mdx muscle cells by antisense oligoribonucleotides." *Hum. Mol. Genet.* 1998. Vol. 5: 1083-1090

Kahn A. "Dix ans de thérapie génique: déceptions et espoirs." *Biofutur* 2000; vol 202: pp.16-20

Ferrari et al. G. "Muscle regeneration by bone marrow-derived myogenic progenitors", *Science* 1998; vol 279, 1528-1530

Gussoni E.et al. "Dystrophin expression in the mdx mouse restored by stem cell transplantation", *Nature* 1999; vol 401: pp. 390-394

Kmiec EB. "Comment réparer les genes", *Biofutur* 1999; vol 195: 32-36

Old L. "L'immunothérapie." *Pour la Science* 1996: No. 229: 106-114

Wang J.S. et al. "Marrow stromal cells for cellular cardiomyoplasty: feasibility and potential clinical advantages." *J. Thorac. Cardiovasc. Surg.* 2000; vol 120: pp. 999-1006

NB: The following contains particularly useful information on current developments in gene therapy: *The Development of Human Gene Therapy.* (1999) (Ed) T. Friedman, Cold Spring Harbor Laboratory Press.

Industry and the human genome

by Mike Furness
and Dr Kenny Pollock

The Human Genome Project (HGP)[1] was initiated in the late 1980s with the belief that such a project would be of great benefit to the understanding of human genetics and diseases, thereby enabling new developments within medicine and the healthcare industry. In reality the impact and potential benefits of the HGP to medicine have probably surpassed initial expectations, and are being realised and applied throughout the whole of the Life Sciences industry.

As the HGP project *per se* approaches completion, the extraction of value from the human genome both from an academic, altruistic perspective (benefits of knowledge to mankind), and from a commercial perspective, is only just beginning to gain significant momentum. It could be said that the current, and projected, value of the human genome can be gauged by the volume of intellectual property currently being filed around genomic data, while in hard cash terms, the estimated value of the genomics industry (products and services) was US$2.2 billion in 1999, growing to an estimated US$8.4 billion by 2004 (Schneider 2000). To this end, the science and application of genomics has developed exponentially over the last ten years, both in concert with, and in parallel to, the HGP and the myriad of other whole genome sequencing projects that have sprung up in recent years, ranging from lower organisms such as bacteria and yeast, to other mammalian species such as rat and mouse (Beeley et al. 2000), and even into the plant kingdom including thale cress *(Arabidopsis thaliana)* (Somerville et al. 1999) and rice *(Oryza sativa),* as recently announced by Syngenta and Myriad (Dickson et al. 2001).

Whilst sequencing provides the "alphabet" of life, genomics seeks to identify, and functionally annotate the basic components of biology starting with the identification of individual genes (the "words") within a stretch of sequence through to the identification and functional annotation of individual proteins (the "sentences") and ultimately to define what role each protein

1.
HGP website: http://www.nhgri.nih.gov/HGP/

plays within a cell, organism or whole animal (the "encyclopaedia" of life). This is not a trivial exercise; it requires the integration of a range of skills from computer science through to applied functional biology. The applications of genomics in industry are wide and varied, but in all cases the key to success lies with the development of appropriate technology platforms and associated data management infrastructures that are capable of genome-wide coverage and have a high throughput for data generation. Complementary to this data generation is the need for huge storage capacity (soon to be hundreds of terabytes) and the capability of analysing the data quickly and in a meaningful way. In this respect the science of bioinformatics (biological informatics) has been growing rapidly to provide the computational tools for handling both numerical data content and associated text information, enabling some biological sense to be made of the numbers (Persidis 1999).

This chapter will aim to discuss some of the technologies currently being applied to genomics, and to give examples of how they are currently having, and may in future have, an impact on the pharmaceutical industry, agrochemical development, food sciences, personal healthcare, and even environmental monitoring.

Genomics and the pharmaceutical industry

Research-based pharmaceutical companies invested US$26.4 billion in research and development in 2000, a 10.1% increase over 1999, which equates to 20% of sales revenues (Pharmaceutical Industry Profile 2000).[1] To maintain growth rate, pharmaceutical and biotech companies must maintain a high research level while controlling corresponding costs. It is estimated that it costs US$500 million to develop a new drug from starting research to reaching the market, and 70% of the cost is accounted for by failures during pre-clinical research and development.[2] Other estimates have shown that, of the drugs that entered phase I clinical trials, an increase in those reaching the market from 23% to 25% would produce an 8% saving in product development costs, or approximately $50 million per product (DiMasi et al. 1991). In addition, it has

1.
See: http://www.
phrma.org/publications/

2.
See: http://www.
phrma.org

been estimated that reductions in the drug approval process by one to two years can reduce costs for generating drugs by 20-23% (DiMasi et al. 1995). However, despite the large costs involved in drug develoment the commercial rewards are substantial for the companies responsible for delivering new medicines. To understand the potential benefits to a pharmaceutical company, we need only look at recent statistics that show that in the US alone an average prescription drug generates US$1.3 million per day in revenue, and in the case of Prilosec® (an anti-ulcer drug), this figures leaps up to approximately US$11.2 million per day (Getz et al. 2000).

Given the potential rewards, the current imperative for the pharmaceutical industry is to identify new, and more appropriate, drug targets against which they can develop newer and better drugs to satisfy unmet medical needs. Implicit in this objective is the need to be more efficient in selecting candidate drugs with improved efficacy, reduced toxicity and less likelihood of adverse effects. The cost savings in developing three successful drugs from ten candidates rather than one successful drug from twenty candidates are obvious. The real benefit, however, is to patients who should be able to receive more effective and safer drugs, not only for currently treatable diseases but also, hopefully, for some that are at present considered untreatable or have low rates of response to therapy.

Figure 1

Breakdown of the drug discovery and development process

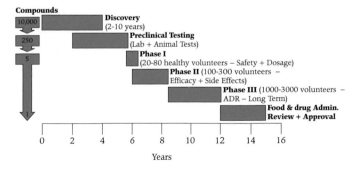

Adapted from PhRMA Report 1998

Lead identification: identifying chemicals which could potentially become drugs.

Lead optimisation: adapting the structure of those chemicals to make them as much like a drug that will be safe and efficacious in the clinic.

Messenger RNA (mRNA): an RNA (ribonucleic acid) that carries the genetic code for a particular protein from the DNA in the cell's nucleus to a ribosome in the cytoplasm and acts as a template, or pattern, for the formation of that protein.

Transcriptome: the thousands of mRNAs specifically expressed in each different cell type.

Phenotype: the physical properties produced by gene expression and environmental influences.

Starting from scratch, the drug discovery process as it stands today in large pharmaceutical companies involves a number of sequential phases as outlined in Figure 1. Briefly, the discovery process is committed to the identification of novel drug targets that contribute to the pathology of specific diseases, and the identification of compounds that interact with those targets, and thereby modify the resulting pathology. This process of chemistry-driven Lead identification* and Lead optimisation* takes two to ten years and results in a drug candidate being progressed to clinical testing. As an alternative approach to medicinal chemistry-driven drug discovery, therapeutics in the form of proteins or antibodies are also being developed, primarily by biotechnology companies. The strategy for exploiting the human genome for protein therapeutics is similar to that used in target identification for small molecules, but the types of proteins that have been identified as drug targets in this context are likely to be different from those identified as therapeutic agents. There are currently around 60 drugs on the market that are proteins and in 1998, 15 of 57 new drugs were proteins or antibodies. It has been estimated that the human genome encodes over ten thousand secreted proteins of which 120 to 280 would have therapeutic potential (Drews 2000). Gene therapy still remains a promise (Friedmann 1996), with many programmes studying the potential benefits. However a death in a recent clinical trial has raised a cautionary note over the prospects of gene therapy from a safety perspective (Smaglik 2000).

In the last few years, the genomics industry has become more focused on delivering tools to enable the rapid identification, validation, and optimisation of drug targets. In living systems, the DNA of which chromosomes are comprised, is transcribed into messenger RNA (mRNA) fragments* which each encode the information required to generate a protein or a peptide (a small protein, or a fragment of a protein). Messenger RNA is therefore the earliest functional output of the genome, and the complex mixture of mRNAs, or transcriptome*, of a cell should reflect the structural and functional phenotype* of that cell. The transcribed mRNAs are then translated into their corresponding protein products, generating the complement of enzymes, structural proteins, etc., which constitute the proteome*.

Using sequence databases in drug discovery

One approach that has proved invaluable to the pharmaceutical industry is the collation of DNA sequence, and the corresponding protein sequence, information in the form of sequence databases. Two approaches to the generation of sequence databases are the most commonly used – sequencing the genome directly, or analysing the expressed mRNA to build an EST (Expressed Sequence Tag) database*. EST sequencing was established in the early 1990s (Adams et al. 1993) and companies such as Incyte (Palo Alto, in California) and Human Genome Sciences, (Rockville, Maryland) have been leaders in generating EST databases as a commercial resource – at present there are over 6.5 million human EST sequences in the Incyte database alone. In addition to these commercial ventures, a public EST database was also established at the Whitehouse Station, New Jersey with funding from Merck (Williamson 1999). When these databases are created, EST sequences are associated with a biological source of origin, such that the relevance of specific sequences can be assessed in the context of particular diseases, or drug treatments. In addition EST databases provide gene-specific links to other public domain sequence databases (GenBank, EMBL, etc.) and literature sources (for example, Medline, OMIM), to facilitate information gathering on those genes already characterised in the public domain (Multiple authors 2001).

The utility of, and value within, sequence databases stems from the application of bioinformatics to the raw data. For example, small fragments of gene sequence can be stitched together *in silico* (electronically, using computers) by using a variety of bioinformatic software tools to identify assemblies of overlapping sequences, and ultimately to identify full length sequences of genes coding for proteins (Schuler 1997; Vasmatzis 1998). EST sequence fragments can also be compared with known sequences that exist in the public databases. EST sequences can also be electronically translated into putative protein sequences that are then subjected to another set of

Proteome: the repertoire of proteins in a cell.

Expressed sequence tag: a short sequence of DNA, representing a gene which has been expressed from the genome.

Homologue:
a gene, or protein, which has similar regions of sequence and/or structure to another gene or protein.

Receptor:
a protein or a group of proteins which transforms the stimuli received from the external environment of the cell into signals that aid the organism to affect behavioural and physiological changes.

Ion channel:
a protein embedded in a cell membrane, serving as a crossing point for the regulated transfer of a specific ion or group of ions across the membrane.

Protease enzymes:
an enzyme that catalyses the splitting of proteins into smaller peptide fractions and amino acids.

bioinformatics tools allowing annotation via sequence alignment/homology searches*. Many novel genes with unknown identity or function are being discovered by this process.

We can start to home in on potential novel drug targets from raw DNA sequences by comparison with known sequences from current drug targets to identify novel homologues. Current estimates are that there are only approximately 500 proteins (Drews 2000) against which drugs have been targeted to date, out of a speculated 5000-10000 targets in the 100000 or more gene transcripts. Most known drug targets fall into a small number of categories, the principal ones being cell surface receptors*, nuclear receptors, ion channels*, and protease enzymes*.

Using expression databases: RNA-based technologies

Following on from sequence databases that have been used to identify the repertoires of individual genes, a number of enabling technologies have been developed to simultaneously monitor the changes in expression of the transcriptome from different diseases or drug treatments. At the mRNA level these technologies include high density microarrays (genes on chips) based on spotted cDNA target sequences (Brown P. et al. 1999; Duggan et al. 1999) which have been commercialised by companies such as Incyte (Palo Alto, California) and Rosetta (Stanford, California), or sequence specific oligonucleotides (Southern et al. 1999; Lipschulz et al. 1999, Kumar et al. 2001) such as those commercialised by Affymetrix (Hayward, California).

Alternative differential expression technologies have been developed commercially including Serial Analysis of Gene Expression (SAGE) (Madden et al. 2000), GeneCalling™ (Rinninger et al. 2000), and READS™ (Prashar et al. 1996) to name but three, that allow identification of large numbers of genes differentially expressed in cells. The advantage of these approaches is that they are capable of identifying the existence, and/or differential expression of, novel genes expressed in very low abundance.

90

Using expression databases: protein-based technologies

Once the genes have been expressed, many are translated into protein. Over the last twenty years, a huge amount of work has been done in developing ways to analyse the proteome rapidly, and as completely as possible. A key development in this respect was the establishment of robust conditions for separating proteins from cells by high resolution 2-dimensional gel electrophoresis (Celis et al. 1999).* Once separated, proteins are stained and located, their co-ordinates on a gel fixed, and the relative abundance of protein at that co-ordinate position estimated across a series of gels. In this way, quantitative changes in a protein are identified. Protein sequence is obtained following excision of the spot and analysis. The derived protein sequence can then be compared with a database of known proteins and some annotation given (Dunn 2000; Page et al. 1999; Celis et al. 2000). To date, the technology allows the identification of up to approximately 2000 proteins from each 2-D gel separation, out of more than 300000 proteins potentially expressed in humans.

The value in protein expression databases will once again come from identifying novel proteins associated with a disease phenotype, with a view to developing these proteins as therapeutics, drug targets or diagnostic markers.* The science of proteomics (the study of the proteome) and its application is technologically still in its infancy, and work is underway to identify a broader, faster way to analyse proteomes, such as the protein chips under development in many laboratories (Davies 2000; Qureshi et al. 2000). While there are advantages to analysing either the proteome or the transcriptome, analysis of both (Celis et al. 2000) allows us a much greater understanding of the biological systems studied.

Target discovery

Having decided upon a favoured technology, an expression database can be created with input from any biological source of interest, whether that be a model cell culture system or tissues removed from humans or animals. In this way experiments can be designed, and data generated, that reflects normal physiology in relation to pathology. A variety of strategies

Gel electrophoresis: a technique for separating molecules of varying sizes in a mixture by moving them through a block of gel by means of an electric field, with smaller molecules moving faster and therefore further than larger ones.

Marker: a specific molecule of known properties (eg, size, sequence, etc.) that is associated with a specific disease or characteristic. This is then used as a reference point, against which we can compare unknown samples, to screen for the disease or characteristic.

Transcriptome profiling:
establishing the pattern, or profile, of genes transcribed seen in a particular cell type.

High throughput screening:
the application of robotics and miniturisation to allow more samples to be tested at faster speeds and lower costs.

Clustering technique:
a mathematical method allowing data to be organized into smaller subgroups, based on similarity. For example, drugs or diseases showing similar patterns of gene changes.

B-cell lymphoma:
a particular subtype of cancer of the lymph gland.

can be used to carry out in-depth analysis of expression data with a view to understanding biology, pharmacology and toxicology at a molecular level, and ultimately to identify candidate drug targets and disease markers. As already discussed in the context of sequence databases, bioinformatics also has a major role in the analysis of transcriptome profiling.*

When dealing with transcriptome profiling data, the results are normalised to allow comparison across numbers of different data sets and across different technology platforms. Following normalisation, the data can be queried (mined) to identify tissue-specific, treatment-specific or disease-specific differences in gene expression within the transcriptome. Although some of these queries are not complex, tens of millions of data points may need to be trawled and analysed simultaneously to extract the pertinent information. From such queries, a candidate list of 50 to 200 genes may be gleaned from a single array of 5 to 10 000 genes (Schiffman et al. 2000). Although a few genes might instantly present themselves as suitable screening targets, amenable to current high throughput screening* (pharmacological target validation), most candidates at this point would require more investigation using other molecular technologies to prioritise them as potential candidates. There may also be instances where a gene is identified with an interesting tissue distribution, and/or a regulated expression pattern in a disease, suggesting that a protein may represent a therapeutic agent or drug target. The use of microarrays for the identification of drug targets is most informative when a candidate gene has some annotation. When a differentially expressed gene is identifiable only as novel EST sequence, more work would be required to elongate and annotate the sequence.

Beyond a simple gene-by-gene analysis of transcriptome profiling data, powerful statistically-based clustering techniques* are now being applied to identify genes with similar patterns of change in expression. Clustering analysis of expression data has been used to identify sub-classes of B-cell lymphomas* (Alizedah et al. 2000) and in a more complex setting to define a classification model for diagnosing different types of leukaemia (Golub et al. 1999). Again through intelligent interrogation of the data, genes will emerge from the data as potential diagnostic markers of disease, alone or in combination with other genes in a cluster, or as candidate drug targets.

Pharmacogenomics

Perhaps what could be considered initially as a sub-discipline of genomics with respect to understanding pathophysiology and target discovery, is the science of pharmacogenomics (Bailey et al. 1999). Pharmacogenomics is the study of a drug's effect on the genome, and is emerging as a powerful strategy to understand disease and characterise biological responses to drugs, from both an efficacy and toxicology perspective, and to identify differences between both normal and diseased tissues. Pharmacogenomics can have an impact on the drug discovery process in a number of areas including target identification, Lead optimisation (understanding and modifying initial chemical structures until they become drugs), and in predictive toxicology. In transcriptome profiling, genes that cluster together in common patterns in different treatment conditions may be characteristic of an individual compound's action, and can be used to characterise the activity of the compound at the gene expression level. Within clusters of interest, individual gene expression patterns can be identified that discriminate between different classes of compounds or even discriminate within a chemical series. Characterising compounds within a chemical series by expression profiling will therefore be a powerful tool for identifying compounds with a preferred profile of activity. However, identifying appropriate activity (efficacy) in a drug development candidate is only half of the story, the other half being safety assessment.

Safety assessment

Once compounds that are potential therapeutics for specific diseases have been identified, one key component of identifying whether or not it will reach human clinical trials, and the market place, is safety assessment. Routinely, compounds will be tested for acute toxicity early in their development cycle, which currently includes the use of a wide range of tests to identify toxicity associated with specific organs (for example, liver toxicity, cardiac toxicity), or more diverse adverse effects (for example, cancer induction, induction of DNA mutations). If these early tests show no adverse effects, the compound will move into more advanced and longer term testing, with the extent and length of the tests being dependent on the extent of clinical trials planned in man. As human trials progress, more long-term studies are also carried out in animals to identify long-term or cumulative effects of the compounds in living organisms.

Much of the testing to date has included a large component of animal work, with most drug filings requiring appropriate toxicity data in one rodent and one non-rodent species. In the last few years, an enormous effort has been placed on developing more rapid, cost-effective, and efficient methods to identify toxicity. For example, it has been shown that only 70% of all clinical toxicities were identified in the current two species tests; this drops to 40% if just rodents are used, and 60% if just non-rodents are used (Olson et al. 2000). So, apart from the ethical and financial issues associated with animal work, there is still a great deal of room for improvement in safety testing procedures used thus far. A recent report summarises the extent of current alternative biological models under assessment by both industrial companies and regulatory bodies (MacGregor et al. 2001).

Some of the current shortcomings in safety assessment procedures could be addressed by applying genomics to certain aspects of toxicology testing. As mentioned earlier, the effects of toxins at the level of the genes can be monitored by transcriptome expression profiling. At present there are toxicology-specific expression databases being created in the public domain (for example, the US National Institute of Environmental Health Sciences (NIEHS) Center for Predictive Toxicology) (Bristol et al. 1996), as commercial products by genomics companies including Incyte (Palo Alto, California), GeneLogic (Gaithesburg, Massachusetts), CuraGen (New Haven, Conneticut), Phase-I Molecular Toxicology (Santa Fe, Arizona), and "in-house" within large pharmaceutical companies such as AstraZeneca (Macclesfield, UK) (Pennie et al. 2000). Having established a reference database, the ultimate aim of this technology would be to identify sets of genes that are associated with a particular type of toxicity, such as liver toxicity, and to then be able to screen new compounds early in development, to remove any which show indications of likely toxicity (Furness et al. 2000). This technology could allow rapid screening of large numbers of compounds in cell models, reducing the number of problematic compounds which progress further, cutting costs and increasing the efficiency of the drug development process. In addition, this would reduce the overall number of animal studies required, as it should minimise the number of compounds that have adverse effects passing into the regulatory safety assessment process.

It is apparent that applying genomics to pre-clinical safety assessment should become increasingly important within the pre-clinical drug development process, to ensure that candidates with the safest profile can be identified prior to entering clinical evaluation. The financial costs of developing what turn out to be toxic drugs are immense. For example data for the last three years shows that ten drugs that have reached the market have since been withdrawn.

Even when drugs have reached the clinic successfully, adverse effects of drugs, or combinations of drugs, can cause many serious problems, and even fatalities, partly as a result of the incomplete understanding of the mechanisms of many drugs. The benefits to the pharmaceutical industry of identifying toxicity profiles and potential adverse effects early on in the development of a drug are therefore enormous.

Pharmacogenetics

Sequence databases are based on the assumption that gene sequences are consistent from one individual to another. However, whilst more than 99.9% of the genome is the same between any two people, variation in 0.1% of the genome is what makes each of us an individual, genetically and phenotypically. Studying the differing genetic sequence information within the genomes of different individuals allows us to begin to both associate small regions of the genome with genetic predisposition to disease, and to identify mutations that may help to explain differences between individual response to disease or drug treatment. On the one hand there are well-known

95

Pharmacogenetics: the study of the effect of changes in a gene's sequence on the activity or function of the protein it encodes.

Nucleotide: fundamental chemical building block of DNA.

Polymorphism: difference in DNA sequence among individuals.

Beta blocker: a substance that blocks the activity of the beta receptors for adrenalin, used especially in the treatment of angina and hypertension.

inheritable monogenetic diseases such as cystic fibrosis, sickle cell anaemia, where there is a single identifiable genetic lesion causing the disease. By contrast the genetic predisposition to major diseases such as cancer, asthma, diabetes, schizophrenia, to name but a few, is multifactorial, complex, and to date, the degree of genetic effect versus environmental is unclear. However, pharmacogenetics* may shed some light on these issues. For example, the identification of single point mutations (single nucleotide* polymorphisms* – SNPs) in genes that can be associated with disease or responses to drugs. Again, this aspect of applied genomics has thrown up a number of high-throughput technologies for identifying SNPs on candidate gene sequences (Campbell et al. 2000) and cataloguing them in the relevant databases (Multiple authors 2001). To date many examples of individual SNPs correlating with disease have been identified – for example, the peroxiome proliferation receptor-gamma in insulin-resistant diabetes (Barroso et al. 1999), the thromboxane A2 receptor in bronchial asthma (Furuta et al. 2000), and the cytokine interleukin-6 in Alzheimer's disease.

In addition to identifying markers of disease, pharmacogenetic analysis of genes involved in the Absorption, Distribution, Metabolism and Excretion (ADME) of drugs is of enormous value. Within the general population there is variation in clinical responsiveness to certain drugs that may be attributable to variation in ADME genes (West et al. 1997; Meyer et al. 1997). Currently, it is known that up to 30% of patients fail to respond to some cholesterol lowering agents (statins), up to 35% do not respond to beta blockers*, and up to 50% have no response to tricyclic antidepressants (Tanne 1998). It has been estimated that as a consequence of individual variation in drug metabolism to compounds, underdosing, overdosing, or missed doses currently cost the United States more than US$100 billion a year in increased hospital admissions, lost productivity, and premature death (Marshall 1997). Variations seen to date can be the result of familial inheritance, ethnic origins, environmental mutations, or a combination of all three factors. Whatever the underlying cause, the application of pharmacogenetics will enable clinical trials to be run more effectively and will ultimately lead to a more effective and appropriate prescription of drugs to an individual.

Structural genomics

Having identified candidate genes of interest as therapeutic targets by carrying out in-depth analysis of expression databases, these potential targets can be progressed to the next level of information – definition of protein structure and function, by comparison with all available protein sequences in the public domain databases (Multiple authors 2001; Tatusov et al. 1997). In the past, structural information on proteins was obtained using a variety of physical methods, including X-ray crystallography.* Although these techniques are now routine, but by no means easy, new approaches are being developed *in silico* to assign structural characteristics to novel unknown proteins (Fagan et al. 2000). The value of structural genomics will come from identifying "druggable" targets by structural comparison with current drug targets, and in predicting the structural consequence of an SNP. For example, if a change in the sequence leads to the protein being made in a shorter form, or generating a different sequence, then the final enzyme it produces will have a different structure, and therefore behave differently. Again the benefits of these technologies stem from informed selection of new targets. In the future, molecular biology-based genomics technologies will link to virtual drug design as is being developed by companies such as Structural Bioinformatics Inc. (San Diego, California) and DeNovo Pharmaceuticals (Cambridge, UK).

Functional genomics

Whilst many candidate genes can be selected from the in-depth analysis of expression data, we can turn the problem around and ask what that gene contributes functionally in a cell or organism. The concept of functional genomics has grown up around transgenic (where a gene is overexpressed appropriately)* and knockout (where a gene is deleted) animals, and characterisation of the consequent phenotype. Many of these animals serve as useful models for drug testing by generating a specific pathology but also to causally link

X-ray crystallography: a method for determining the 3-D structure of a protein.

Overexpressed gene: one which is transcribed in much higher quantities than normal, which usually leads to an increase in the protein for which that gene codes.

selected gene expression with pathology. Generating trans-genic/knockout animals is a slow process, taking months in mice. As a quicker, higher throughput alternative to mammalian species, in recent years lower animal species, including yeast *(Saccharomyces cerivisiae),* nematodes *(Caenorhabditis elegans),* and fruit flies, *(Drosophila melanogaster)* have been used, through a combination of genetics and genomics, to identify genes that are essential to a variety of key biological processes. There is immense value in modelling biological processes in model organisms to identify genes in critical biological processes as was shown by the identification of the CED3 cell death gene in *C. elegans,* homologues of which play a vital role in controlling cell survival in all mammalian systems studied to date (Yuan et al. 1993). Taking this approach further, there are now many functional genomics-based target/drug discovery platforms (for example, Exelexsis (San Francisco, California), PPD (Hayward, California), Galapagos (Leiden, the Netherlands), Cellomics (Pittsburgh, Pennsylvania)) based in lower organisms or mammalian cell culture systems that rely on phenotypic selection. In these systems, candidate genes are introduced to, or deleted from, the host cells using classical genetics or by gene targeting. Cells with an altered phenotype are isolated, and the gene responsible for the defect identified. In this way, potential drug targets can be characterised to identify their functional role. In some cases, knowing the function of the potential drug target will allow us to also identify a number of drugs that may be expected to have some activity against the function of the target.

Other industries affected by genomics

So far we have talked only about the impact genomics will have on the pharmaceutical and biotechnology industries. Many of these applications also apply to other industries, and we will aim to identify examples of these in the next section.

Veterinary science and animal health

Much of the genomics technology that was focused on generating the human genome sequence, has also been running in parallel in a number of species, whether they be model experimental animals, such as rats and mice, or commercially valuable animals, such as cattle, pigs and chickens. This sequence information allows us to better understand the diseases of those animals, the effects of parasites and external agents on the animals, and potentially, the effects of veterinary medicines on animals. In the case of commercially important animals, there is the added prospect of genetically manipulating animals to create, for example, larger or leaner cattle or sheep, and to speed up the historic process of cross-breeding to enhance an animal's commercially valuable attributes.

Obviously, in the area of veterinary medicine, there is an overlap with human medicine, with some drugs being used across man and many other different species (for example, antibacterial drugs, sedatives, hormone treatments). All of these drugs need to pass through a similar set of safety and regulatory studies as those that are used for human pharmaceuticals.

The food industry

The first obvious application of genomics is in the identification of genome sequences of crop plants. Identification of the genes within these genomes that generate "positive attributes" in plants for example, will allow us to generate crops with greater resistance to drought, with enhanced resistance to parasitic attack or infection, or with increased nutritional value (Maleck et al. 2000). There has even been work done in both commercial and academic laboratories to identify ways of enhancing the nutritional value of foods, such as production of crops like Golden Rice, which produces vitamin A not normally endemic to the plant species (Schiermeier 2001), potentially adding to the nutritional value of the rice in Third World communities, where rice is a staple food, but vitamin deficiency is a major health problem.

This leads into the new growth area of so-called "functional foods". These are foods that may enhance certain physiological parameters in the consumer (German et al. 1999). One well-publicised recent example is Benecol® (Johnson & Johnson) which uses phytosterols (a type of lipid or fat found in plants) in the manufacture of a number of dairy products or dairy sub-

Goiter: enlarged thyroid gland sometimes resulting from iodine deficiency.

stitutes, which have been claimed to show reduced blood cholesterol levels in some consumers. Currently there is extensive debate in this area, and also about the extent to which safety testing needs to be applied to these foods – do substances such as Benecol® count as additives, or should they undergo more rigorous testing? (Jacobson et al. 1999). There is no doubt that historically, additives to foods have shown positive health benefits, such as iodised salt (sea salt) curing iodine deficiency goiter,* and grain products fortified with oat bran which may help reduce the risk of cardiac disease. There is still a need for caution in regulatory policy surrounding the inclusion of food "supplements" needed to alleviate public concerns, and to ensure the safety of consumers from charlatans, who may try to cash in on a market which is estimated to be worth US$12 billion a year in the United States alone (Brown D. 1999). The application of genomic technologies to these products may help us better understand their effects on consumers, and become integrated into the safety testing procedures that already apply to foodstuffs.

Personal healthcare

The personal healthcare industry covers a broad range of areas, ranging from the production of shampoo, perfume, sun creams and toothpaste, to name but a few examples. All of these products need to be developed in the context of their application. For example, an understanding of skin structure and function is key to developing appropriate skin treatment products. We need to understand what pathological changes take place in skin when it becomes sun-damaged or is exposed to anti-inflammatory drugs, or how oral pathogens are affected by the components of toothpaste. All of these fundamental questions can be addressed in the same ways as we try to understand disease biology, applying all the genomics technologies discussed previously. In addition, all these products, and/or their components, must undergo safety testing before being allowed onto the market, so we have the potential impacts of genomics on safety assessment that can also be applied here.

There is also a degree of overlap between the pharmaceutical, food and personal healthcare product industries, as all need to test their products for safety in humans. Some of this synergy has led to alliances between these apparently disparate industries to address a number of the key issues in safety assessment (Basketter et al. 2000), as already discussed earlier in this chapter.

Agrochemicals

The agrochemical industry could almost be considered in the same light as the pharmaceutical industry, in that the goal of the development of an agrochemical is usually to destroy pests and infections relating to crops. In this case we need to understand the biology of the plants, and those of the pathogens or parasites which infest them. We can then use genomics technologies to identify genes which are present in the pathogen or the parasite, but not in the plants, to make selective toxins against the infection or infestation. One caveat is obviously that we must also understand the genomes and genetics of many other species, especially humans, to check that these genes are exclusive to the parasite, and no other organism. For example, targeting proteins responsible for metabolism of lanosterol (a type of fat) in the cell walls of fungi, has led to the formulation of effective "fungi-specific" drugs such as fluconazole (Kowalsky 1991).

Environmental and ecological monitoring

While the discussion so far has been on the potential impacts of genome technologies in industry, as this is the main focus of this chapter, it is worth mentioning that these technologies have the capacity to be applied equally to so-called "green" issues, such as environmental and ecological monitoring. This can be most obviously seen in the application of transcriptome expression profiling to identify the effect of different levels of pollutants on humans, animals and plants. In this situation, comparing the expression profiles of an organism from a polluting or toxic agent, with those of a normal, untreated same organism, and profiles previously seen with known agents, we can begin to identify the likely causative agent, and at the same time, start to understand the effect the agent is having on the biological system as a whole. For example, studies on yeast treated with a herbicide have been carried out to try and identify whole organism effects, using microarrays to monitor changes in the transcriptome profile (Jia et al. 2000).

The field of genomics has grown exponentially over the last decade, and the sequencing of the human genome by the Human Genome Project can be seen as one of the recent pinnacles in human endeavour. In discussing the use of genomics in industry, we have mentioned but a few of the private and public organisations and companies that are currently contributing to this field. However, from the examples given, it is clear that the knowledge that is beginning to appear from the Human Genome Project is likely to have an enormous impact on a wide range of industries, bringing many positive benefits to the everyday lives of people throughout the world.

References

Adams, M.D., et al. "3 400 new expressed sequence tags identify diversity of transcripts in human brain", *Nat. Genet.,* 4, (1993), p. 256

Alizedah, A.A., et al. "Distinct types of diffuse large B-cell lymphoma identified by gene expression profiling", *Nature,* 403, (2000) p. 503

Bailey, D.S., et al. "Pharmacogenomics – it's not just pharmacogenetics", *Curr. Opin. Biotech.,* 9, (1999), p. 595

Barroso. I., et al. "Dominant negative mutations in human PPAR-γ associated with severe insulin resistance, diabetes mellitus and hypertension." *Nature,* 402, (1999), p. 880

Basketter, D.A., et al. "Use of the local lymph node assay for the estimation of relative contact allergenic potency", *Contact Dermatitis,* 42, (2000), p. 344

Beeley, L.J., et al. "The impact of genomics on drug discovery", *Prog. Med. Chem.,* 37, (2000), p. 1

Bhojak, T.J., et al. "Genetic polymorphisms in cathepsin-D and interleukin-6 genes and the risk of Alzheimer's disease", *Neurosci. Lett,* 288, (2000), p. 21

Bristol, D.W., et al. "The NIEHS Predictive-Toxicology Evaluation Project", *Envir. Health Persp.,* 104, suppl. 5, (1996), p. 1001

Brown, D. "FDA Commissioner enters war over words", *Washington Post,* A31, 26 March 1999.

Brown, P., et al. "Exploring the new world of the genome with DNA microarrays", *Nature Genet.,* 21 supplement, (1999), p. 33

Campbell D.A., et al. "Making drug discovery a SN(i)P", *DDT,* 5, (2000), p. 388

Celis J.E., et al. "2D protein electrophoresis: can it be perfected?", *Curr. Opin. Biotech.,* 10, (1999), pp. 16-21

Celis, J.E., et al. "Gene expression profiling: monitoring transcription and translation products using DNA microarrays and proteomics", *FEBS Lett.,* 480, (2000), p. 2

Chouchane, L. "A repeat polymorphism in interleukin-4 gene is highly associated with specific clinical phenotypes of asthma", *Int. Arch. Allergy Immunol.,* 120, (1999), p. 50

Davies, H.A. "The protein chip system from Ciphergen", *J. Mol. Med.,* 78, (2000), B29

Dickson, D., et al. "Commercial Sector Scores Success with Whole Rice Genome", *Nature,* 409, (2001), p. 551

DiMasi, J.A., et al. "Cost of innovation in the pharmaceutical industry", *J. Health Econ.,* 10, (1991), p. 107

DiMasi, J.A., et al. "Research and development costs for new drugs by therapeutic category: a study of the US pharmaceutical industry", *Pharmacoeconomics,* 7, (1995) p. 152

Drews, J.J. "Drug Discovery: A Historical Perspective", *Science,* 287, (2000), p. 1960

Duggan. D.J., et al. "Expression profiling using cDNA microarrays", *Nature Genet.,* 21 supplement, (1999), p. 10

Dunn, M.J. "Studying heart disease using the proteomic approach", *DDT,* 5, (2000), p. 76

Fagan R., et al. "Bioinformatics, target discovery and the pharmaceutical/biotechnology industry", *Curr. Opin. Mol. Ther.,* 2, (2000), p. 655

Fisher, D.A., et al. "Isolation and Characterization of PDE9A, a Novel Human cGMP-specific Phosphodiesterase", *J. Biol. Chem.,* 273, (1998), p. 15559

Friedmann, T. "Human Gene Therapy – an immature genie, but certainly out of the bottle", *Nature Med.,* 2, (1996), p. 144

Furness, L.M., et al. "Expression databases – resources for pharmacogenomic R&D." *Pharmacogenomics,* 1, (2000), p. 281

Furuta, A.M., et al. "Association studies of 33 single nucleotide polymorphisms (SNP)s in 29 candidate genes for bronchial asthma: positive association with a T924C polymorphism in the thromboxane-A2 receptor gene", *Hum. Genet.,* 106, (2000), p. 440

German, B., et al. "The development of functional foods: lessons from the gut", *TIBTECH,* 17, (1999), p. 92

Getz, K.A., et al. "Breaking the development speed barrier: assessing successful practices of the fastest drug developing companies." *DIJ,* 34, (2000), p. 725

Golub, T.R., et al. "Molecular classification of cancer; class discovery and class prediction by gene expression monitoring", *Science,* 286, (1999), p. 531

Jacobson, M.F., et al. "Functional Foods: health boon or quackery?" *BMJ,* 319, (1999), p. 205

Jia, M.H., et al. "Global expression profiling of yeast treated with an inhibitor of amino acid biosynthesis, sulfometron methyl", *Physiol. Genomics,* 3, (2000), p. 83

Kowalsky, S.F., et al. "Fluconazole: a new antifungal agent", *Clin. Pharm.,* 10, (1991), p. 179

Kumar, A., et al. "Chemical nanoprinting: a novel method for fabricating DNA microchips", *Nuc. Acids Res.,* 29, (2001), p. 2e

Lipshulz. R.J., et al. "High density synthetic oligonucleotide arrays", *Nature Genet.,* 21 supplement, (1999), p20

MacGregor, J.T., et al. "*In vitro* tissue models in risk assessment: report of a consensus-building workshop", *Toxicol. Sci.,* 59, (2001), p. 17

Madden, S.L., et al. "Serial analysis of gene expression: from gene discovery to target identification", *DDT,* 5, (2000), p. 425

Maleck, K., et al. "The transcriptome of Arabidopsis thaliana during systemic acquired resistance", *Nature Genet.,* 26, (2000), p. 403

Marshall, A. "Getting the right drug into the right patient", *Nature Biotech.,* 15, (1997), p. 1249

Meyer, U.A., et al. "Molecular mechanisms of genetic polymorphisms of drug metabolism", *Ann Rev. Pharmacol. Toxicol.,* 37, (1997), p. 269

Multiple authors, "The 2001 Database Issue", *Nuc. Acids Res.,* 29 (2001)

Olson, H., et al. "Concordance of the toxicity of pharmaceuticals in humans and in animals", *Regul. Toxicol. Pharmacol.,* 32, (2000), p. 56

Page, M.J., et al. "Proteomic definition of normal human luminal and myoepithelial breast cells purified from reduction mammoplasties", *Proc. Nat. Acad. Sci.,* USA 96, (1999), p. 12589

Pennie, W.D., et al. "The Principles and Practice of Toxicogenomics: Applications and Opportunities", *Toxicol. Sci.,* 54, (2000), p. 277

Persidis, A. "Bioinformatics", *Nature Biotech.* 17, (1999), p. 828

Prashar, Y., et al. "Analysis of differential gene expression by display of 3' end restriction fragments of cDNAs", *Proc. Nat. Acad. Sci.,* USA 93, (1996), p. 659

Qureshi A.Q., et al. "Large-scale functional analysis using peptide or protein arrays", *Nature Biotech.,* 18, (2000), p. 393

Rinninger, J.A., et al. "Differential gene expression technologies for identifying surrogate markers of drug efficacy and toxicity", *DDT,* 5, (2000), p. 560

Schiermeier, Q. "Designer rice to combat diet deficiencies makes its debut", *Nature,* 409, (2001), p. 551

Shiffman, D., et al. "Large scale gene expression analysis of cholesterol-loaded macrophages", *J. Biol. Chem.,* 275, (2000), p. 37324

Schneider, J. "From sequence to sales: the genomics payoff." *R&D Directions,* April, (2000), pp. 38-48

Schuler, G.D. "Pieces of the puzzle: expressed sequence tags and the catalog of human genes", *J. Mol. Med.,* 75, (1997), p. 694

Smaglik, P. "Congress gets tough with gene therapy", *Nature,* 403, (2000), p. 583

Somerville, C., et al. "Plant Functional Genomics", *Science,* 285, (1999), p. 380

Southern, E., et al. "Molecular interactions on microarrays", *Nature Genet.,* 21 supplement, (1999), p. 5

Strader, C.D., et al. "Molecular approaches to the discovery of new treatments for obesity", *Curr. Opin. Chem. Biol.,* 1, (1997), p. 204

Tanne, J.H. "The new world in designer drugs." *BMJ,* 316, (1998), p. 1930

Tatusov R.L., et al. "A genomic perspective on protein families", *Science,* 278, (1997), p. 631

Vasmatzis,G., et al. "Discovery of three genes specifically expressed in human prostate by expressed sequence tag database analysis", *Proc. Nat. Acad. Sci.,* USA 95, (1998), p. 300

West, W.W., et al. "Interpatient variability: genetic predisposition and other genetic factors", *J. Clin. Pharmacol.,* 37, (1997), p. 635

Williamson, A.R. "The Merck Gene Index project", *DDT,* 4, (1999), p. 115

Yuan, J., et al. "The *C. elegans* cell death gene CED3 encodes a protein similar to mammalian interleukin-1beta-converting enzyme", *Cell,* 75, (1993), p. 641

The human genome: individual property or common heritage?

by Professor Bartha Maria Knoppers

The notion of ownership of human genetic material has been and continues to be a contested topic. Indeed, the use of human genetic material and the information it contains has both reproductive, therapeutic, and commercial value. The same is true for plants and animals, but symbolically and politically the notion of ownership of human genetic material has drawn more attention.

There has been and continues to be a legal divide between property and persons but the possibility of patenting human genes is subsuming this issue. Furthermore the notion of individual ownership itself is becoming largely secondary to that of consent and control, irrespective of legal characterisation. Finally, the notion of individual "ownership" in the context of human genetic research and DNA banking is being accompanied by the interests of communities and populations in benefit-sharing. Perhaps, with increasing understanding of human genetics and of the legal basis of the notion of common heritage of humanity, we can finally move the debate away from ownership to developing procedures that enhance and recognise both personal control and that of communities and populations.

Persons and property: the classical divide

The concept of property refers not only to rights over material objects but also to intangible things such as intellectual property. The rights that property confers can be limited to a simple right of use, or extend to the right to sell and so to profit. It is common parlance to say that the human body is outside the realm of sale and commerce, as exemplified in the prohibition of slavery.

No legal case has explicitly held that genetic material and the information it contains is either property or part of the person.[1] Leaving aside human organs, gametes and embryos to

1.
Moore v. Regents of University of California, 271 Cal. Rptr 146 (Cal. S.C. 1990). This case did not settle the issue of property but rather held that the issue was one of informed consent.

concentrate on the issue of human genes, the importance of such a characterisation has not however escaped political notice be it international, regional or national.

Representative of the international position is the 1997 Unesco Universal Declaration on the Human Genome and Human Rights which maintains that "[t]he human genome in its natural state shall not give rise to financial gain" (Article 4).[1] This position then does not characterise human genetic material but rather upholds the principle of non-commercialisation. Likewise, the European Convention on Human Rights and Biomedicine of the same year also insisted that "the human body and its parts shall not, as such, give rise to financial gain" (Article 21).[2] But these pronouncements had more to do with an attempt to influence the ongoing patenting debate than to consecrate the status of human genetic material as "person".

Indeed, in the reality of research, of pharmacogenomics and of DNA banking, the consent process and the possibility of eventual commercialisation are the same under both the person or property approaches. While DNA banking is not the topic of this chapter, it is interesting to note that even in countries such as France where the gene in its natural state is described as non-patrimonial in the bioethics laws of 1994 or in those American states that have adopted the Genetic Privacy Act imposing a property characterisation, the consent options provided for research participants for DNA banking are the same. Moreover, the participant is also simply informed that the research may eventually be commercialised and lead to intellectual property such as patents but that they will not reap any personal economic benefit (Knoppers 1999). Hence, the debate has focused away from the issue of its qualification as person or property to that of patenting.

Patenting

Since 1992, the Human Genome Organisation (HUGO) has maintained the non-patentability of human gene sequences of unknown utility. Yet debate and controversy was such that the 1998 European Community Directive concerning the Legal Protection of Biotechnology Inventions[3] was a decade in the

1.
Unesco, Universal Declaration on the Human Genome and Human Rights, see *Appendix II* for website.

2.
The full text of the Convention for the Protection of Human Rights and Dignity of the Human Being (4 April 1997) with regard to the Application of Biology and Medicine is available at: http://book.coe.fr/conv/en/ui/frm/f164-e.htm

3.
European Union Directive 98/44/EC of the European Parliament and of the Council of 6 July 1998 on the legal protection of biotechnological interventions, L 213 Official Journal, (30 July 1998), pp. 0013-0021, can be accessed online at: http://europa.eu.int/eur-lex/en/lif/dat/1998/en_398L0044.html

making. This directive reiterated the principle of non-commercialisation and thus of the non-patentability of DNA sequences as follows: "The human body, at the various stages of its formation and development, and the simple discovery of one of its elements, including the sequence or partial sequence of a gene, cannot constitute patentable inventions" (Article 5.1). According to Article 6, even a patentable invention can be excluded if contrary to public order and morality. Furthermore, the recital mandates that the person whose DNA is to be banked must consent to eventual commercialisation according to national law (recital 26).

Even though this directive should seemingly have closed the debate, two political events serve to demonstrate the continued viability of the issue of patentability. In March 2000, both Tony Blair, Prime Minister of the United Kingdom and Bill Clinton, then President of the United States announced that gene sequences of unknown utility were part of the public domain (see below, Common heritage). Moreover, in June 2000, the National Consultative Ethics Committee of France,[1] a country which had been part of the decade of negotiations leading to the European Union Directive turned against the directive.

The French National Ethics Committee took the position that the mere isolation or cloning of a gene could not be an invention but constituted a simple discovery. This was already common knowledge since industrial (utility) application would still have to be demonstrated. The committee also alleged that the directive ran against the 1994 bioethics laws even though the law was in force when France participated in the debate leading up to the adoption of the directive. These alleged ambiguities led the Ethics Committee to question the adherence of France to the Directive.

The position of the French Ethics Committee is all the more surprising considering the fact that the Directive itself contains further explanations and delimitations on the possible interpretation of Articles 5.1[2] and 5.2.[3] For example, Article 5.3 requires that the industrial application be clearly described, and according to recital 24, which provides further guidance for interpretation, if a sequence is used to produce a protein,

1.
National Consultative Ethics Committee, Opinion No. 64 on a preliminary draft law incorporating transposition into the Code of Intellectual Property, of European Union Directive 98/44/EC, dated July 6 1998, on the legal protection of biotechnological inventions, (8 June, 2000), http://www.ccne-ethique.org/english/start.htm

2.
European Union Directive 98/44/EC, *op. cit.,* note 5, Article 5.1

3.
European Union Directive 98/44/EC, Article 5.2: "An element isolated from the human body or otherwise produced by means of a technical process, including the sequence or partial sequence of a gene, may constitute a patentable invention, even if the structure of that element is identical to that of a natural element".

the precise protein or its function needs to be disclosed. Article 7 gives the European Group on Ethics in Science and Technology the responsibility of evaluating all ethical issues in biotechnology.

On 5 January 2001 the US Patent and Trademark Office announced that it would turn down patents unless applicants state at least one "specific, credible and substantial use for the gene"[1]. The key objective is to deter applicants who sequence genes or gene fragments of only limited or often speculative use. Thus, it is no longer possible to obtain a patent simply by saying that a newly sequenced gene could be used as a molecular probe for detecting that same gene in humans. Now, the applicant must show why finding a gene is important, for example, by linking its presence with a disease.

It bears noting that on 16 June 1999 the administration of the European Union amended the implementation requirements of the directive. Thus, it is currently being applied to patent applications in the fifteen countries of the European Union[2]. In short, greater certainty is being achieved on the issue of gene patenting, by the increasingly restrictive interpretations of demonstrable industrial application (utility).

Now that patentability criteria are being clarified and tightened up, does that mean that the issue of ownership will go away and that we can concentrate on the conditions for consent and control in DNA sampling? This is not likely, for even the notion of individual, informed consent is being increasingly accompanied by the claims of communities and populations to participate in the consent process and benefit from any eventual profits.

Populations and benefit-sharing

On 9 April 2000 the HUGO (Human Genome Organisation) Ethics Committee released its Statement on Benefit-Sharing.[3] It maintained three fundamental arguments in favour of benefit-sharing. First, we share 99.9% of our genetic makeup with all other humans. In the interest of human solidarity, we owe each other a share in common goods, such as health. Second,

1.
Department of Commerce Patent and Trademark Office, "Utility Examination Guidelines", (5 January 2001) 66 *Federal Register* 1092.

2.
Administrative Council of the European Patent Organisation, Decision of the Administrative Council of 16 June 1999 amending the Implementing Regulations to the European Patent Convention, (16 June 1999), h t t p : / / w w w . european-patent-office.org/epo/ca/e/ 16_06_99_impl_e.htm

3.
Human Genome Organisation (HUGO), *Statement on Benefit-Sharing*, (9 April 2000), http:// www.gene.ucl.ac.uk/ hugo/benefit.html

starting with Hugo Grotius's law of the sea in the seventeenth century and proceeding to international law governing air and space in the twentieth century, such global resources have been viewed as common, equitably and peacefully available to all humanity, and protected in the interests of future generations. International law may therefore set a precedent for regarding the human genome as a common heritage. Third, when there is a vast difference in power between an organisation carrying out research and the people providing material for that research, and when the organisation stands to make a substantial profit (albeit taking the risk of investment), concerns about exploitation arise that benefit-sharing can address. The HUGO Ethics Commitee maintained that considerations of justice required action to meet basic healthcare needs.

Already in 1996, its Statement on the Principled Conduct of Genetic Research[1] had proposed community consultation and participation in the framing of research that affected or involved whole populations and sub-populations. This statement also suggested that there be recognition of such participation (but not inducement through payment) through technology transfer, local training, immediate medical care, or contribution to a humanitarian cause or healthcare infrastructure. The 2001 Benefit-Sharing Statement reiterated this goal and further specified that in addition to thanking participants and acting in accordance with community values and preferences, 1-3% of net profits (if any) be provided for the needs of the general population.

To give a few examples, Iceland has opted for a return to the population (free-of-charge) of any products or drugs discovered using their database (Gulcher and Stefansson 2000). Tonga has obtained technology transfer, similar promises of free drugs, and annual research funding for its Ministry of Health from revenues generated from any discoveries that are commercialised (Anon 2000). Like the current situation in Iceland however, exclusive access to the database is held by one company. Estonia's national DNA bank has chosen not to anonymise the samples and so can return medical information to the population (Estonian Parliament 2000). No monopoly of the database is foreseen.

1.
HUGO, *Statement on the Principled Conduct of Genetic Research*, (21 March 1996), http://www.gene.ucl.ac.uk/hugo/conduct.htm

This concept of benefit-sharing moves the debate away from ownership to considerations of equity and justice. Nevertheless, too much emphasis on individual or even community ownership can have the undesired effect of underscoring and so supporting commodification. Indeed, the concept of benefit-sharing can only serve to reinforce the notion of the human genome itself as the common heritage of humanity.

Common heritage

Intrinsically opposed to state ownership but not to state sovereignty, this concept as mentioned earlier, dates back to the seventeenth century and has been used to protect the air and the sea in the name of the common interests of the international community. Its core elements are as follows:

1. Any area designated as a common heritage is not to be appropriated.
2. An international authority should manage all use of the areas and their resources.
3. Any benefits arising from exploitation of the areas and their resources will be shared equitably.
4. The areas and resources are to be used only in peaceful ways.
5. The areas and their resources are to be protected and preserved for the benefit of present and future generations.

To quote the Chair of Unesco's International Bioethics Committee's Legal Commission: "The very fact that the human genome is proclaimed as the common heritage of humanity reaffirms the rights of each individual over his genetic heritage, [that is] – as something individual, intransferable and which cannot be repudiated – is of interest to humanity in its entirety as a subject of law. Humanity, the legally organized international community, protects this heritage and ensures that it cannot be appropriated by any individual or collective body, whether this be a state, [a] nation of people" (Espiell 1997).

Unfortunately, lack of knowledge of these legal criteria and political wrangling led to this concept being watered down by government representatives in Article 1 of the 1997 Unesco Universal Declaration on the Human Genome and Human Rights which now considers the human genome "[i]n a symbolic sense [to be] the heritage of humanity"[1].

Indeed, the International Bioethics Committee had embraced the "common heritage of humanity" concept, but certain government representatives designated to study and approve the committee's final draft declaration understood the common heritage concept as mandating possible appropriation by international conglomerates and thus a risk to state sovereignty. Others disliked the community aspect. Ironically, other members of the Bioethics Committee, fearful of possible state sovereignty, preferred to protect the human genome at the level of the individual. Finally, the French translation of heritage as "patrimony" also created difficulties since it would be seen as having an economic meaning. Hence, the adoption of the expression "symbolic of the heritage of humanity".

Certain conclusions can be drawn on the issue of the status of human genetic material:

1. At the level of the individual, neither its legal status as person or property, nor the prohibition of individual financial gain for its use, affects its potential for patentability.

2. The patentability criteria have been considerably clarified and tightened up.

3. The emergence of the concept of benefit-sharing in the areas of debate on the human genome and patentability may well serve (if supported and enforced), to mitigate any untoward effects of patenting.

4. Its status in the collective sense, at the level of of the human genome, has not yet been legally accepted as being in the common or public domain, though it has been recognised as the heritage of "humanity". This latter failure weakens the public policy arguments in the patent debate.

The time is ripe to move forward beyond the reification or sanctification of human DNA. We need to examine and harmonise the conditions of consent and control at the level of

1.
See *Appendix II* for website.

individuals within countries as well as ensuring international surveillance at the level of the common human genome. Classical debates create classical divides. Compulsory licensing of gene patents in the healthcare sector, or clarifying and expanding the traditional research exemption in the health sector under patent law are but two new issues meriting discussion.[1] Such discussion will probably have greater impact than the current sterile polemic surrounding "ownership".

References

Anonymous, "Autralians hunt for Tonga's disease genes", (November 30, 2000) 408 *Nature* 508.

Espiell, H.Gros, "Unesco's Draft Universal Declaration on the Human Genome and Human Rights", (1997) 7 *Law & Hum Gen Rev* 133.

Estonian Parliament, Human Genes Research Act, (December 13, 2000), *RT I 2000,* 104, 685.

Gulcher, J.R Stefansson, K. "The Icelandic healthcare database and informed consent", (2000), 342 *New Engl J Med* 1827-30.

Knoppers, Bartha Maria, "Status, sale and patenting of human genetic material: an international survey", (1999) 22 *Nature Genetics* 23

1.
Patent law already has specific provisions for protecting the rights of researchers through the research exemption provision. These provisions allow experimentation with the invention without infringing the patent.

At the frontiers of humanity

by Professor Dr Jens Reich

In June 2000, at a press conference held at the White House, Washington DC, attended by US President Clinton, UK Prime Minister Tony Blair and the ambassadors of several countries, the spokesmen of two scientific consortia announced that a draft version of the human genome had been submitted to the computers. Although this draft was actually very far from being finalised, the event was accompanied by a series of triumphant epithets uttered by the two leaders as well as by James Watson, who fifty years earlier had published his famous DNA double helix, the paradigmatic model of the genomic era to come.

Shotgun technique: Technique by which large individual DNA fragments of known position are subjected to shotgun sequencing, that is, they are shredded into small fragments that are sequenced, and then reassembled on the basis of sequence overlaps.

The announcement of the draft was compared to the landing of human beings on the moon; it was characterised as the most important map ever constructed – a new Book of Life that would replace the Holy Scriptures. Comments were also made as to the effect such an achievement would have on the pride and self-confidence of humankind; it was seen as a climax of biological evolution, and reference was also made to the high aspirations held out for applying genome knowledge to anthropology and medicine. The politicians present emphasised the importance of protecting human rights and personal information in the new era of genomic analysis that they predicted to be impending.

The celebrations may have been somewhat premature, in particular as the sequence text had so far been obtained only in the form of many millions of hashed text pieces that still needed to be assembled to create a full "library". Six months later, in February 2001, after frantic work on the early version, the draft assembly of the human genome was made public. One of the research consortia involved was publicly funded by international academic projects, and the other was set up by a commercial company, Celera Genomics, which had collected an enormous amount of venture capital in order to launch a brute-force attack on the sequence of the human genome by the so-called shotgun technique*.

In order to obtain and assemble a full set of text pieces, Celera Genomics integrated the sequencing and mapping achievements of the International Human Genome Sequencing Consortium, so that scientifically – though neither psychologically nor commercially – one could describe the results as a joint achievement. There are still considerable gaps in the sequencing of the human genome, and their completion will take some time, as these parts are particularly difficult to decipher. Nevertheless, more than 90% of the genome is now well aligned in data files, and it is worth reflecting on the impact of this information.

The human genome is encoded in the 46 chromosomes of every human cell, in the form of DNA. The data files contain collections of texts, written in an alphabet of four letters, each of which symbolises a distinct chemical entity (nucleotide). Thus we have a chemically-encoded text. The total of this text amounts to 6 billion letters, approximately 50% of which are redundant, being a copy of the other half. As had been earlier predicted, it has now been firmly established that 99.9% of the human genome is the same for the whole species, while only 0.1% varies among the members of the species. In total our genome contains about 6 million letters that may be different from one individual to the next. The individual part is of great importance for individual genetics and medical application, but from the point of view of biological evolution the invariant part is more important.

Humans and other species

Our closest relatives in the animal kingdom are the anthropoid apes, closest of all probably the chimpanzee *(Pan troglodytes).* Its genome has so far not been sequenced in great detail, but pilot studies as well as experimental assessments on global identity have revealed that about 98.5% of the chimp's genome, when counted as text sequence, is identical to that of *Homo sapiens.* This is a staggering similarity, even if visits to the zoo have already brought to light the fact that we are pretty similar to these animals in terms of anatomical structure and behaviour. A difference of 1.5% is amazing, because the difference even between humans can be much higher in terms of percentage, when we compare males and females; since the X-chromosome (of which women have two, and men only one with a Y-chromosome in place of the second X-chromosome) contains about 5% of the information in our genome, we could say that men and women differ by 5%, that is to say more than humans and chimps of the same sex. Admittedly, the chimpanzee *(Pan troglodytes)* has 48 rather than our 46 chromosomes, which means that the DNA text is organised in different packages.

Of course such summary calculations are only approximate and general; it is the detailed differences which matter much more. The globin gene, for example, which contains the prescription for the synthesis of the red blood pigment essential for life because of its capacity to transport oxygen, is almost identical in the two species. Whereas in the case of genes that are crucial, for instance for brain development, the results might be totally different – we simply do not know at present. Since coding segments of the genome, that is those that carry vital information, fill only 2-3% of the whole text, these differences may be extremely significant; this would be in keeping with what we suspect. Admittedly, the human as well as the ape brain "expresses" a considerable fraction of its genes, that is the brain uses it for making proteins. We know this from protein fingerprinting, which indicates that there are over 10 000 proteins in the nerve cells. It will be of crucial importance for anthropology to study these nerve-specific regulatory and structural genes in detail.

Thus the similarities between man and chimp are striking and revealing, but they are not as surprising as the similarities between the gene content of the human genome and the gene content to be found in other species, which we are in the habit of calling "more primitive". When quantifying the number of genes, that is the number of text sections that are blueprints for the synthesis of proteins[1] as specific body material, it is more informative to take into account the number of encoded genes instead of the sum of all the letters of the genome. The number of genes has been estimated in a number of species apart from *Homo sapiens,* including the mouse *(Mus musculus),*

1.
See Appendix I.

the fruit fly *(Drosophila melanogaster)*, the nematode worm *(Caenorhabditis elegans),* baker's yeast *(Saccharomyces cerevisiae),* and the plant thale cress *(Arabidopsis thaliana).* Before I mention specific numbers in this context, I should point out that we still do not have all the information on most of the genes in any one of these species. To have complete information about a gene, we need to know how the gene product itself is characterised, ie to have information on the purified protein that it encodes. Establishing this is a formidable task that has been carried out for only about 10% of all gene products, even for baker's yeast about which we have the most advanced knowledge so far.

Nevertheless, the number of genes, and a rough classification of them, can be obtained through statistical estimates. Analysis of the genes that we are confident have been accurately established shows that encoding texts may be distinguished from non-encoding ones by "linguistic" properties; letter usage, frequency of letter combinations and letter clusters, the presence of certain unequivocal signals, etc. Such statistical analysis is comparable to the study of texts in information coding theory: it is possible to identify whether an encoded text is written in English or German, and even whether it is of, say, military or non-military content, even if one is completely unaware of the meaning of individual phrases or words. There are computer algorithms that have identified all the proteins encoded in the yeast genome. Direct chemical analysis has revealed that the majority of these computer estimates turn out to be correct. For the human genome, the identification of the genes is much more difficult, and until very recently, estimates varied wildly, with the different computational and biochemical methods predicting fundamentally different numbers. However, for chromosomes 21 and 22 (which are, admittedly, two of the smaller ones of our set) international consortia have presented an annotation of the genome text, and have made an estimate of the corresponding number of genes, which, as far as we know, should be very close to the actual figure. Furthermore, these estimates may be applied to the whole genome with a certain level of confidence. The most accurate figures produced by both sequencing and annotation consortia today estimate that the human genome contains about 32 000 statistically well-established coding genes, to which a further 10 000 might be added at a later stage, when we are more confident of the analyses. This is a number which is considerably below earlier estimates, which sometimes exceeded 100 000; this could be explained, however, by subtle technical reasons which affected the accuracy of earlier estimates.

Now 32 000 plus some thousands. What does this mean? The comparison with the estimate of gene numbers in the other more fully studied species is revealing:

- the nematode worm *(C. elegans)* contains 19 000 genes;
- the fruit fly *(D. melanogaster)* 14 000 genes;
- baker's yeast *(S.cerevisiae)* 6 000;
- thale cress *(A. thaliana)* 25 000 genes.

The comparison is surprising. Are we not infinitely more complex in our physical appearances, biochemical diversity and brain structure than, for example, the tiny nematode worm *(C. elegans)* (which is barely two millimetres long, and which has a strictly fixed number of microscopically visible body cells, namely 971, compared to a rough estimate of 100 trillion body cells in an adult human)? How can we achieve this level of complexity with such a small number of genes?

The answer to such questions is twofold. On the one hand it seems obvious: yes, we are not that different. The structure of body cells, of their nucleus and membranes, and the basic biochemistry of nutritional and energy metabolism has been maintained over billions of years of divergent evolution. There is a considerable number of genes in the worm or the fly and in our human genome which are so similar that only the specialist can distinguish between them in the test tube or on the computer screen. The fundamental building bricks and the basic biochemistry and biophysics of all living matter are the same. They are so alike that we may consume, much to our benefit, even the most remotely related species (aesthetic disinclination towards eating an ugly worm notwithstanding). The basic biological makeup is so similar, and this has been brilliantly confirmed by the hitherto impossible analysis of the informational elements of life encoded in the genome, that we may, in future, confidently screen thousands of biologically active compounds of all sorts for their inhibiting effect on budding yeast cells in order to obtain candidates for the treatment of human cancer: growth regulation is similar enough in both cases to warrant usage of one as an arguably limited model of the other. The gene which triggers the development of the eye in the fruit fly is still present in the human genome, albeit of course with a sequence that has changed somewhat during the hundreds of millions years of divergent evolution since the last common ancestor. Not only is it present, it is also functional. It has even been shown that this gene is involved in the development of the human eye, which is structurally and anatomically so different from the insect's eye.

On the other hand, however, different species, even ones which are quite closely related, are worlds apart, if not in the quantitative, then certainly in the qualitative aspects of developmental and biological specificity. It is the interaction rather than the sheer number of gene products or the chemical nature of the building bricks that makes species so different in quality. We know, for instance, that more than one hundred different gene products act as signals in the triggering of cell growth and cell division – a process called the "cell cycle" in all living creatures. Still, the enzymes, receptor proteins and modifying factors involved here are, in principle, the same everywhere, but only in the elementary sense that two different pieces of music played on a piano are the same. (I recognise that it is a piece of classical music long before I distinguish any specific structure to the piece or any subtlety in its performance). By analogy, fruit fly larva and the human embryo have the same segmental structure, but one uses it for an external shell of chitin and the other for an internal vertebral column to obtain a stabilised body structure. A third example, the worm for instance, does not form a stabilised body structure in spite of its segmental body plan. Small wonder that many components, genes included, are the same in many species while crucial other ones are subtly specific. It is the task of a future generation of biologists to unravel the tangle, and to distinguish, hopefully for our benefit, the elements which we share and those in which we differ, from the enormously complex interaction that emanates from a very similar start, that is, the fertilised egg cell, to the millions of different body and brain structures within the animal kingdom.

Xenotransplantation

Another immediate practical consequence of the new category of knowledge about the human condition concerns our attitudes towards xenotransplantation, which subsumes, in theory but not yet in practice, all the possible types of tissue and organ transplantation from foreign-species (*xenos* is the paradoxical Greek word for foreign, as well as host and guest) suppliers to human receivers. With the high degree of genome similarity between, say, man and pig, one might argue that the moral outrage about organ transplantation is obsolete, and we might even approach moral exoneration by replacing human donors (which are required to be dead, or dying, in the case of vital organs) by animal donors like the pig whose internal organs fit the functional and size demands of human acceptors very closely. In general, Europeans do not hesitate to assimilate pig tissues by eating their parts, why not incorporate them intact?

There are technical arguments against this; rather paradoxically these rely on both the surprising easiness with which the species barrier can be crossed and on its continued impenetrability. The immune barrier between man and any mammal, the high percentage similarity of the genome text notwithstanding, is so high that the receiving organism reacts with an extremely fierce rejection, which can be overcome only by "humanising" the donor's immune system, that is by creating a chimera between animal and man. On the other hand, the species and immune barriers are so easily penetrable that a virus that is well tolerated by the donor species might transform into a deadly plague when it is transferred to human beings by way of a transplanted organ.

But such arguments are purely pragmatic. They might be tackled by genomic reconstruction. There is a deeper apprehension, in my opinion, which precludes xenotransplantation of whole organs. It is again spiritual, not biologically pragmatic. I am convinced that we are not allowed to make hybrid creatures, for example a cross between goats and sheep. I think that each species has its own evolutionary fixed mode of existence, and we are not entitled to surpass these boundaries of historical fact. We are obliged to allow every animal, even if it is exclusively raised to serve our needs, its species-specific way of life. This rule is violated by the production of chimeras. *A fortiori,* we are not allowed to create a chimera between man and any animal; this would violate our own dignity as human beings. I am convinced that in future the only categorically acceptable form of transplantation will be the autotransplantation of material grown from our body, from our own adult stem cells. This is the only moral way, ultimately, of avoiding the use of other creatures for our individual purposes. And this is the most practical way of avoiding the immune rejection of grafts. I should add in fairness that my position is not unanimously shared by all biologists.

Hybridity

There is another puzzling aspect in the similarities and dissimilarities between *Homo sapiens* and "lower" or "higher" animals and plants. This has to do with the species barrier. Species, even if they are rather similar in their appearances, are believed to be categorically different. In Judaeo-Christian belief systems the Holy Scriptures decree that God created the species fundamentally different, and that the attempt of humans to mate with animals is a perversion. More than that, it is futile: the decisive criterion of taxonomy (the discipline that classifies animals and plants) to distinguish a species barrier between otherwise similar creatures is the inability to produce fertile progeny. There will never be, despite all the fantasies of gifted writers, and despite the aforementioned 98.5 % identity of the genome, a naturally begotten hybrid between a human and a chimpanzee. All sorts of hybrids created between horses and donkeys, for example, are infertile in the following generation. Between man and chimpanzee it is quite clear: the text is nearly the same, but its organisation at chromosome level is so different that conjunction of germ cells cannot function. The only form of chimera between different species that has so far been obtained is at cellular level: in cell cultures. It is the developmental programme of the whole organism rather than some elements that is "specific", that is species-making. And most people react with instinctive disgust against any attempt to create chimera between, say, domestic animals such as sheep and goats. It is my opinion that we should leave the evolved species as diverse and distinct as they are. Preservation of species barrier is also one of the most convincing critical arguments, to my mind, against the genetic modification of crops by the introduction of certain foreign genes.

In the light of such culturally internalised biological facts, it is surprising and a real challenge to learn what intensive traffic has taken place between *Homo sapiens* (or his ancestors) and even the most remote species. The human genome contains hundreds of thousands of genes (or, at least, gene remnants) of viral origin, and even hundreds of bacterial genes – only the specialist can distinguish the sequence of a true bacterial gene from its homologue in our genome, which renders very convincing the hypothesis that we have adopted them by interspecies transfer. That such a transfer is possible is testified by the HIV virus, which, according to the most probable theory, has jumped from some monkey species onto *Homo sapiens* as a possible host. The virus has also proved its ability to inscribe itself into the human genome, namely into that of white blood cell precursors in the bone marrow. It inscribes its own text into our genome and waits for better times. From time to time the HIV genome becomes expressed and creates a bout of infection, which, when repeated in the long term, overstresses and ruins the immune system: Aids is the final outbreak of the disease. So HIV could be seen to demonstrate both interspecies traffic and invasion into the genome.

The many virus genes that we inherit are all very probably disabled. According to all circumstantial evidence, they were acquired many million generations ago, have spread over the whole species, and consist mostly of fragments of intact virus genomes. However, the bacterial genes are probably functional, and the recent retrovirus transcriptions like those from HIV, certainly are. No one can rule out the possibility that the sleeping or dead genes in our genome cannot be revitalised by the transfer of gene segments from independent sources (recombination of microbial genomes is quite common, for example). So we are living with a considerable number of lodgers, or corpses of lodgers, which are living quite comfortably and in part even to our benefit: bacteria, after all, are unsurpassed masters of detoxification, and it may be extremely profitable to our genome to acquire certain biochemical tools from them (as we do on a different level by amply using their inventions in industrial biotechnology).

The sequencing of the human genome has confirmed what has long been known from different, less impressive sources and facts: the unity and diversity of the biosphere. We have a genome of similar organisation and with a similar number of function modules (genes and others), we consume billions of gene molecules with every bit of food; most of it, but not all, is disintegrated inside the gut. What escapes digestion may be absorbed and enter our venous systems and hence the liver. In very rare cases, foreign genes may enter some of our cells, and still more rarely, may become integrated into the genome. If the material is able to invade the germ cells (which are well protected anyway) then it may become part of the genome of progeny. And random distribution over thousands of generations may finally spread it into the whole species. These descriptions show that the barriers are not absolute, but they are high. Changing our genome as established by evolution is not an everyday matter. But it can happen. And it certainly has happened. This eventuality is not merely a remote historical event of no practical concern, nor just a playground for absent-minded biologists. We are well aware of the limited exclusiveness of the species barrier when considering, for instance, the use of animal organs for transplantation purposes. The possibility that this may liberate a hitherto harmless, but potentially harmful micro-organism is sufficiently well-founded to warrant very careful proceedings in this direction.

An aside to these speculative studies of human genomics is that gene transfer may also occur the other way round. After all, we have thousands of bacterial species as commensals in our gut. And if, after death, we are not burnt, then an army of micro-organisms and multicellular creatures will disintegrate our corpse. Might it occur that one day these beasts benefit from what we have learned, the complexity of gene regulation for instance? I think this is improbable, but not impossible.

Some people may regard the findings on the human genome as the final metaphysical insult after Charles Darwin taught us that we were descended from apes and Sigmund Freud asserted that we are not in command of our consciousness. Now we even appear to be no better than worms and flies. Such a depressing impression is flawed, however. We are all made from the same matter, but we are still very, very different. We may use the first fact to our benefit and take heart from the latter one. After all, human beings are more than just reflex machines, even in strict biological terms. Our main difference in category, however, is metabiological, metaphysical and spiritual: on this issue the biologist can simply contribute interesting reflections, as the substance of the matter lies elsewhere.

Conclusion

by Professor Jean-François Mattei

The decoding of the human genome indisputably represents a major upheaval not only in medical progress but also in the understanding of the human being.

The necessary ethical debate

Ending these writings on the human genome, I consider myself doubly fortunate to be a doctor with many years of practice in medical genetics and also a political office-holder helping to devise the regulatory framework for a scientific discipline which carries high ethical risks. As a member of the Council of Europe Parliamentary Assembly participating in the work of the Steering Committee on Bioethics, I am well aware how influential cultural differences can be in the drafting of international instruments. Not, of course, in the sense of some people having a nobler vision than others of humankind, or of dividing the good from the bad; rather, the force of our convictions often makes us stand firmly on values which, being part of our own identity, are impossible to efface. Time-honoured principles are not easy to renounce on the ground that researchers have supposedly found new evidence capable of altering outlooks on life, death, destiny or difference.

Anyone confronted with the novel situations born of the scientific, or in this case genetic, revolution seeks a solution closely conforming to his mental picture of humankind. A dialogue between the individual and his consciousness ensues, revolving around his philosophical, moral and religious landmarks and resulting in individual convictions which are necessary to guide personal choices at certain stages of life. Individual convictions deserve the deepest respect, and everything must be done to ensure that nobody is under compulsion to break with them. Nor, however, can it be overlooked that each individual choice is impossible to visualise in the abstract, divorced from reality and aloof from fellow-men and time itself.

On the contrary, each personal choice has inevitable consequences reaching beyond the self to others and from the present into the future. Otherness and time are imperative considerations from which the concept of responsibility necessarily follows. Superimposed on the ethic of individual conviction is the ethic of responsibility that prompts the definition of common rules enabling us to live together as individuals and exist as a society.

The difficulty of this approach should not be underestimated, as all who are or have been faced with such situations know full well. There is no ethical choice without moral tension. Receptiveness and respect are necessary, but so are persuasion and courage in pursuing a debate which, however patient, is never dispassionate. Obstacles abound. Plainly, there is no question of legitimising any moral order that would privilege political society to be the arbiter of good and evil. It can be seen clearly enough how much misery and intolerance is caused in various places by various brands of fundamentalist and integrist movements which offend against the very idea of democracy. Neither is it to be implied that all solutions are of equal worth, obliterating the landmarks. The moral values vital to the harmonious public life of all human societies must therefore be restored where they are lacking.

There is no question either of confusing the two distinct registers of legality and morality, even though they may be interwoven across indistinct dividing lines.

Finally, striking the proper balance between respect for the individual and collective necessity is something of an impossible challenge which must be taken up nonetheless.

All these points have bearing on knowledge of the human genome.

From progress to selection

The first impulse is of course justifiable pride in scientific progress. It is always an uplifting achievement to push back the frontiers of knowledge and open new avenues for medicine and care of patients. Identification of genes and of their disease or disability-causing mutations makes it possible to effect accurate diagnoses, inform the families and prevent the occurrence or recurrence of often severe and as yet incurable complaints. But in counterpoint to this we glimpse lurking danger in the general application of antenatal diagnoses leading to elimination of impaired embryos or foetuses, significantly encouraged by echography moreover.

Without any intention of reviving the eugenic creed with its dire historical connotations, the fact nevertheless remains that selection of unborn children is an obvious temptation and in my view constitutes a real hazard. By straying in that direction, our societies would radically alter their outlook on the disabled and the more vulnerable. It would be so much simpler, cheaper and less troublesome if elimination prevented their ever being born, than to take care of them for a lifetime which is steadily growing longer. Compassion would lend a hand in making it more and more common to judge some lives not worth living ("wrongful life"), to the extent of being considered an actual injury subject to compensation. Clear confirmation of this societal trend can be seen in a recent decision by the French Court of Cassation, the country's highest judicial authority.

Regarding identification of genes that predispose to a given disease, Professor Jean Dausset's contribution showed the significant progress which accrues from this form of predictive medicine in allowing a foreseeable illness to be prevented, managed, and treated if necessary. However, there has been too ready an inclination to let the idea of a real biological destiny develop from a few sparse examples; a notion of man being the prisoner of his genes, as it were. This encapsulates the entire thesis of determinism and of course challenges the humanity of mankind, denied all freedom and consequently absolved of responsibility. This purely genetic conception of humankind must naturally be refuted and, since most of the diseases in question are of polygenic or multifactorial origin, I much prefer the term "presumptive medicine" as more clearly implying predisposition and not ineluctable prediction. This drift into determinism has been further accentuated by the purported discovery of genes associated with certain types of behaviour such as violence and homosexuality. There again, a danger is visible in the new approach to certain criminal behaviour: the offender would be regarded more as a sufferer, the sentence as a prescription, and re-offending as a medical failure.

In any case it is plain that given the recent advances of genetics our societies, having already attempted to combat social inequalities, will have to make provision for genetic inequalities. It is indeed unacceptable, for the sake of essential solidarity, that human beings should have to bear social discrimination on account of defective genes which they have presumably inherited.

Genes and patents

The eminent specialist Professor Knoppers has put forward altogether cogent and firmly founded arguments on the topic of possible ownership of genes. Legally, there is nothing to criticise, and moreover her viewpoint is quite frequently echoed by patent law specialists and some industrialists. Biotechnologies do in fact foreshadow great strides in diagnosis, immunisation and treatment methods. They must be assisted, and I am convinced that they will soon permit the healing of many, so far incurable, diseases. Using normal genes to replace defective ones or to produce human therapeutic proteins; developing the replacement of ageing and/or abnormal cells with normal young cells: these are wonderful strategies that absolutely require investment.

Nevertheless, the biological revolution should be attended by a legal one in my view, because the advent of the biotechnologies which are to revolutionise tomorrow's medicine creates a wholly unprecedented situation. Use of living matter for the benefit of the living world cannot but raise novel and crucial questions. How should living matter, that decidedly original raw material, be perceived? How is to be procured, at what price and under what conditions? All these questions raise speculation as to the soundness of gene patentability, further considering that patents on living organisms are nothing new;

on this issue, one sometimes has the impression of arriving when the battle is over.

I need not reiterate the distinction between discovery and invention, although it ought not to be forgotten; a patent protects an invention and is inapplicable to a discovery. We should firstly consider arguments of an ontological nature. Man, as a person endowed with dignity, cannot be traded, which is why organs, tissues and cells are not marketable and do not come within the terms of the normal market. The gene, as the smallest component part of the human being, cannot to my mind be treated otherwise than the human being and enter directly or indirectly into a business rationale. Another such argument concerns the collective dimension which has been approached in discussing genes. Human genes are indeed a common heritage of mankind, the joint property of humanity in so far as they are handed down from generation to generation and shared among members of families and populations. Since they belong to everyone, nobody can assume the right to claim sole ownership of them, even temporarily by means of a patent. This reasoning might furthermore extend to all living matter without being confined strictly to human genes.

But there are still other arguments against the patentability of genes.

Research arguments come first. Extension of the patenting system to DNA sequences was admittedly justified by the desire to encourage research investment, but today the adverse effects of the system are being experienced. Researchers do indeed require unrestricted access to knowledge for basic and applied research alike. In the background is the entire problem of public and private research as complementary processes with differing objectives but a common goal in the quest for knowledge. I might add that the genes of orphan diseases, having no market likely to afford a worthwhile return on investment, cannot be dependent on patents in the commercial and industrial sense.

Now for some arguments relating to industry itself. Is it reasonable for anyone to want to claim a monopoly over a natural, particularly a human, raw material? Let us imagine that if a total genome is patented, industrialists who acquired a patent in respect of a small percent of genes would nevertheless, logically, be under an obligation to pay for access to all the rest. The real competitive economy is the one that seeks to perfect techniques, the added value of inventing the process, for obtaining products as reliably, economically and effectively as possible from raw material. Clearly patents must play a part in the whole process of innovation and generation of products. It is impossible to imagine that a commercial company would be given a monopoly of a gene sequence for the purpose of devising a competitive method. This is not in the best interests of patients, or industry itself, which has more to gain from its know-how in devising methodologies, products and therapies than from holding a patent on a gene sequence. After all, nobody has ever wanted to patent electrons, but that has not prevented the thriving development of the entire electronics industry. Furthermore, polygeny is so frequent and biological phenomena are so complex that it would very quickly become a difficult management proposition to use large numbers of genes under different patents.

Recently, the marketing of tests for predisposition to breast cancer highlighted another adverse effect of gene patenting. This example illustrates the distortions caused when firms appropriate human genes which they have helped discover. Appropriation of human genes in fact warps access to genetic testing, and the doomsday scenario predicts widespread testing without genetic counselling or preventive policy. It is suggested that the secretive management of industrial property which prevails today be replaced by very open licensing systems.

Finally, it must be acknowledged that the future is forfeited through confiscation of knowledge by patenting. The joint HGP sequencing project associates just six of all the world's countries to varying degrees; this state of affairs appears unacceptable in the long run for all the countries excluded, and especially the developing countries.

Quite apart from the progress achieved in the realm of scientific knowledge, one is bound to conclude on the many questions raised by knowledge of the human genome. Quite plainly, and so much the better, genetics will provide responses to a whole series of diseases and afford a better understanding of biological evolution mechanisms, major human migrations, and population genetics. However, genetics will not explain the remarkable complexity observed in man who, amazingly, possesses only twice as many genes as the nematode worm. Thus, study of genetics already opens up extraordinary new fields of investigation and indefinitely postpones the revelation of man's innermost secrets. For it is evident that man cannot amount to the sum total of his genes or his double spiral of DNA. Is it then permissible to speculate that the stuff of man's humanity does not lie in his material being?

Appendix I – Some key concepts

Cells

All human beings are made up of cells. The simplest organisms (such as bacteria) comprise only one cell without a nucleus and reproduce by dividing into two. All the cells of multicellular organisms (human beings, animals, plants) have a nucleus and reproduce in a much more complex way: the nucleus and certain elements contained in the cell, the mitochondria, replicate. The cells of multicellular organisms join together and develop in a specific way to form muscles, organs, leaves, roots, etc.

Chromosomes

The nucleus of each cell contains chromosomes which are visible only when the cell divides. The number of chromosomes of each species is constant and characteristic of that species: 1 for bacteria, 14 for peas, 38 for cats, 48 for chimpanzees, 58 for chickens, 94 for goldfish. Human beings have 46 chromosomes, in 23 pairs, one chromosome of each pair being inherited from the mother and the second from the father.

Somatic cells, that is cells of the body of an organism (with the exception of the gametes or sex cells) are diploid, meaning they have a paired set of each chromosome. The germ (sex) cells, ovum and spermatozoon, are haploid, that is they have only a single set. During fertilisation the 23 pairs of chromosomes are re-established when the nuclei of the ovum and the spermatozoon combine. There is just one pair of chromosomes, the sex chromosomes, which is not identical in males – the X and Y chromosome. Females have two X chromosomes.

DNA

Each chromosome is a long thread of DNA (deoxyribonucleic acid), the molecule that contains all the information necessary to produce a living being. It therefore holds the keys of heredity. DNA is a compilation of nucleotides which in turn are

Chromosome

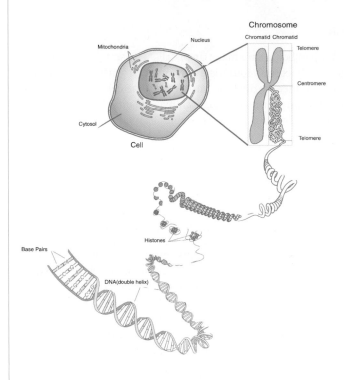

National Institutes of Health

National Human Genome Research Institute
Division of Intramural Research

formed from three elements: phosphate, a sugar and a variable base: adenine (A), guanine (G), thymine (T) or cytosine (C). Each nucleotide is named after its base – A, G, T or C.

In 1953, two researchers, James Watson and Francis Crick, discovered that the DNA molecule was composed of two complementary strands forming a double helix. The two strands are held together by weak chemical bonds between the four bases. Base pairs form only between G and C and between A and T.

DNA replication

This complementary linkage between the bases means that the base sequence of each single strand can be deduced from that of its partner. It can therefore be duplicated (including in the laboratory). This is what happens during cell division and is termed "DNA replication". It is this process which transmits the genetic information from generation to generation.

When cells divide, enzymes separate the two strands between which there is only a weak bond. Other enzymes, DNA polymerases, attach themselves to each strand and bind a free T nucleotide, floating in a cell, to an A nucleotide; it does the same with a free G nucleotide to a C nucleotide, and so on, thereby forming a second strand. In humans there are 46 DNA double helix strands to be copied as there are 23 pairs of chromosomes. On average, this process takes eight hours.

Genes

The combination of nucleotides along the DNA form sentences or "sequences" in a language of four letters: A, T, C and G. These sequences have a particular role to play in the synthesis of proteins, the molecules required for cells to live. These sequences are the linear arrays of genes along the chromosomes.

Human beings have around 30 000 paired genes (one transmitted by the mother and the other by the father). Genes contain all the instructions (codes) required to produce proteins. Our biological functions, our physical appearance, the organisation of our body etc. all depend on one or more genes. Changes to the DNA, termed "mutations", can adversely affect the development or correct functioning of an organism with varying degrees of severity. These mutations, which can also result from environment-related factors, may cause hereditary diseases and cancers.

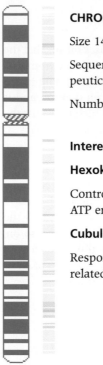

CHROMOSOME 10

Size 146 Mb

Sequenced by: the Sanger Centre; Genome Therapeutics Corporation; IMBB/FORTH

Number of genes: 1440

Interesting genes:

Hexokinase 1

Controls first step in conversion of glucose into ATP energy.

Cubulin

Responsible for the uptake of vitamin B12 and related molecules in the intestine

Proteins

Proteins are macromolecules composed of one or more chains of amino acids. Examples of proteins are the keratin of the skin and hair, collagen in the tendons, haemoglobin in the blood, enzymes and insulin. The number of amino acids in proteins is extremely variable. There are 20 different amino acids in all living beings; each one is coded in triplets (codons). This makes up the genetic code (discovered in 1963), which applies to all living beings.

As there are three successive bases chosen among the four DNA bases, there are 64 (43) possible codons, which means that more than one codon "encodes" the same amino acid. For example, the codons GCA, GCC and GCG encode alanine, the codons CAA, CAG, CAT encode valine, etc. Of all the possible codons, three (ATT, ATC and ACT) do not encode any amino acid: they are "nonsense" or stop codons which signify the end of protein synthesis and, therefore, the end of the gene.

Because the genetic code is universal, it is possible to make human genes function in other organisms, animal or even plant. The study of genetic diseases and their treatment by gene therapy is based on a knowledge of the sequences of mutated genes which are the cause of such diseases.

From gene to protein

How does a cell get a protein out of a gene?

Genes contain information encoded in the nitrogenous bases – adenine (A), thymine (T), guanine (G) and cytosine (C) – in a sequence of DNA.

In the nucleus, where most cellular DNA is contained, this message of As, Ts, Gs and Cs is transcribed into a strand of RNA. RNA contains information within a similar set of bases as is found in DNA, with one important exception: it relies on uracil (U) instead of thymine (T). (These two bases differ only by a single chemical modification.)

So, does that mean that a strand of RNA is identical to the DNA from which it was transcribed? Not exactly.

The newly transcribed RNA strand is complementary to the original DNA sequence. This complementarity is based on the natural pairing of certain combinations of bases – A with U, G with C – but not others. Thus, a DNA sequence of TACTTGGCA would be transcribed into an RNA sequence of AUGAACCGU.

This strand of RNA – called messenger RNA (mRNA) – undergoes a process of "editing" in which certain segments are removed and it leaves the nucleus.

Following this modification, mRNA docks at a cellular organelle called a ribosome. This is where tRNA molecules deliver amino acids corresponding to the triplet-encoded information on the mRNA.

Thus the process of translation is begun – the conversion of the RNA message into a polypeptide, the next stage in protein synthesis.

Translation of the information contained in mRNA to a polypeptide, or string of amino acids, can be compared to translation of a sentence from English to Spanish: the meaning has not changed, but the "language" has.

A polypeptide then undergoes a series of modifications, which may include addition of chemical groups such as sugars or combination with other polypeptides, to form a "mature" functional protein.

Source: GeneLetter (http://www.geneletter.org)

Mendel's laws

In 1866, Gregor Johann Mendel, a Moravian monk discovered the laws of heredity. When he cross-pollinated peas with yellow seeds and peas with green seeds, he noticed that all the peas obtained had green seeds. In contrast, self-pollination of first generation peas gave a result of 25% peas with yellow seeds.

He concluded that one form (green) of a trait (colour) was dominant in the first generation and the second form (yellow) appeared in the second generation. Today we know that the dominant (green) and recessive (yellow) forms of a trait are linked to the dominant and recessive genes (alleles). For example, an individual who has inherited two recessive alleles (one from the mother and one from the father) will express the recessive characteristic. An individual who has inherited both a dominant and a recessive allele will express the dominant trait. This explains the differences between individuals in the same species and the fact that traits from previous generations can be inherited.

□ ○

GG gg

First generation

□ □ □ □

Gg Gg Gg Gg

Second generation

□ □ □ ○

GG Gg Gg gg

G = dominant gene

g = recessive gene

Appendix II – Websites

General information

AFM
http://www.afm-france.org

GeneLetter
http://www.geneletter.com

Genome links
http://www.well.ox.ac.uk/~johnb/genomic.html

Genome web
http://www.hgmp.mrc.ac.uk/GenomeWeb

(US) HGP website
http://www.ornl.gov/hgmis

National Human Genome Research Institute
http://www.nhgri.nih.gov/HGP

Nature magazine
http://www.nature.com/nature
http://www.nature.com/genomics/human/

Science magazine
http://www.sciencemag.org

Wellcome Trust
http://www.wellcome.ac.uk

Companies and research

Celera Genomics
http://www.celera.com/genomics/genomics.cfm

Fondation Jean Dausset (CEPH)
http://www.cephb.fr

Estonian Genome Foundation
http://www.genomics.ee

Généthon
http://www.genethon.fr

GeneSage
http://www.genesage.com/company/people.html

Human Genome Sciences
http://www.hgsi.com

Incyte genomics
http://www.incyte.com

Myriad genetics
http://www.myriad.com

NCBI genome sequencing
http://www.ncbi.nlm.nih.gov/genome/seq

The Institute for genomics research (Tigr)
http://www.tigr.org

Conventions, directives and treaties

Administrative Council of the European Patent Organisation, Decision of the Administrative Council of 16 June 1999 amending the Implementing Regulations to the European Patent Convention
http://www.european-patent-office.org/epo/ca/e/16_06_99_impl_e.htm

Council of Europe Parliamentary Assembly Recommendation 1512
http://stars.coe.int/ta/ta01/EREC1512.HTM

Council of Europe's Convention for the Protection of Human Rights and Dignity of the Human Being with regard to the Application of Biology and Medicine
http://book.coe.fr/conv/en/ui/frm/f164-e.htm

Report on biotechnologies by the Council of Europe Parliamentary Assembly's Committee on Science and Technology
http://stars.coe.int/doc/doc00/edoc8738.htm

Directive 98/44/EC of the European Union of 6 July 1998 on the legal protection of biotechnological interventions
http://europa.eu.int/eur-lex/en/lif/dat/1998/en_398L0044.html

French National Consultative Bioethics Committee Opinion on a preliminary draft law incorporating transposition into the Code of Intellectual Property of Directive 98/44/EC, dated 6 July on the legal protection of biotechnological inventions:
http://www.ccne-ethique.org/english/avis/a_064.htm#deb

Human Genome Organisation (HUGO), Statement on Benefit-Sharing, (9 April 2000)
http://www.gene.ucl.ac.uk/hugo/benefit.html

HUGO, Statement on the Principled Conduct of Genetic Research, (21 March 1996)
http://www.gene.ucl.ac.uk/hugo/conduct.htm

Unesco's Declaration on the Human Genome and Human Rights (1997)
http://www.unesco.org/human_rights/hrbc.htm

United Nations Convention on Biological Diversity
http://www.biodiv.org/convention/articles.asp

Genome sequence data

Celera Genomics
http://public.celera.com/index.cfm

United States National Center for Biotechnology Information
http://www.ncbi.nlm.nih.gov/entrez/query.fcgi?db=Genome

University of California at Santa Cruz
http://genome.ucsc.edu

Industry and the genome

Pharmaceutical Industry Profile 2000
http://www.phrma.org/publications

Pharmaceutical Research and Manufacturers of America (PhRMA)
http://www.phrma.org

Other science resources on the net

Association for Science Education
http://www.ase.org.uk/index.html

New Scientist Planet Science
http://www.newscientist.com

NFER
http://www.nfer.ac.uk/nfer2.htm

Sciencenet
http://www.sciencenet.org.uk

The Why Files
http://whyfiles.org

Sales agents for publications of the Council of Europe
Agents de vente des publications du Conseil de l'Europe

AUSTRALIA/AUSTRALIE
Hunter Publications, 58A, Gipps Street
AUS-3066 COLLINGWOOD, Victoria
Tel.: (61) 3 9417 5361
Fax: (61) 3 9419 7154
E-mail: Sales@hunter-pubs.com.au
http://www.hunter-pubs.com.au

AUSTRIA/AUTRICHE
Gerold und Co., Weihburggasse 26
A-1010 WIEN
Tel.: (43) 1 533 5014
Fax: (43) 1 533 5014 18
E-mail: buch@gerold.telecom.at
http://www.gerold.at

BELGIUM/BELGIQUE
La Librairie européenne SA
50, avenue A. Jonnart
B-1200 BRUXELLES 20
Tel.: (32) 2 734 0281
Fax: (32) 2 735 0860
E-mail: info@libeurop.be
http://www.libeurop.bee

Jean de Lannoy
202, avenue du Roi
B-1190 BRUXELLES
Tel.: (32) 2 538 4308
Fax: (32) 2 538 0841
E-mail: jean.de.lannoy@euronet.be
http://www.jean-de-lannoy.be

CANADA
Renouf Publishing Company Limited
5369 Chemin Canotek Road
CDN-OTTAWA, Ontario, K1J 9J3
Tel.: (1) 613 745 2665
Fax: (1) 613 745 7660
E-mail: order.dept@renoufbooks.com
http://www.renoufbooks.com

**CZECH REPUBLIC/
RÉPUBLIQUE TCHÈQUE**
USIS, Publication Service
Havelkova 22
CZ-130 00 PRAHA 3
Tel.: (420) 2 210 02 111
Fax: (420) 2 242 21 1484
E-mail: posta@uvis.cz
http://www.usiscr.cz/

DENMARK/DANEMARK
Swets Blackwell A/S
Jagtvej 169 B, 2 Sal
DK-2100 KOBENHAVN O
Tel.: (45) 39 15 79 15
Fax: (45) 39 15 79 10
E-mail: info@dk.swetsblackwell.com

FINLAND/FINLANDE
Akateeminen Kirjakauppa
Keskuskatu 1, PO Box 218
FIN-00381 HELSINKI
Tel.: (358) 9 121 41
Fax: (358) 9 121 4450
E-mail: akatilaus@stockmann.fi
http://www.akatilaus.akateeminen.com

FRANCE
La Documentation française
124 rue H. Barbusse
93308 Aubervilliers Cedex
Tel.: (33) 01 40 15 70 00
Fax: (33) 01 40 15 68 00
E-mail: commandes.vel@ladocfrancaise.gouv.f
http://www.ladocfrancaise.gouv.fr

GERMANY/ALLEMAGNE
UNO Verlag
Am Hofgarten 10
D-53113 BONN
Tel.: (49) 2 28 94 90 20
Fax: (49) 2 28 94 90 222
E-mail: unoverlag@aol.com
http://www.uno-verlag.de

GREECE/GRÈCE
Librairie Kauffmann
Mavrokordatou 9
GR-ATHINAI 106 78
Tel.: (30) 1 38 29 283
Fax: (30) 1 38 33 967

HUNGARY/HONGRIE
Euro Info Service
Hungexpo Europa Kozpont ter 1
H-1101 BUDAPEST
Tel.: (361) 264 8270
Fax: (361) 264 8271
E-mail: euroinfo@euroinfo.hu
http://www.euroinfo.hu

ITALY/ITALIE
Libreria Commissionaria Sansoni
Via Duca di Calabria 1/1, CP 552
I-50125 FIRENZE
Tel.: (39) 556 4831
Fax: (39) 556 41257
E-mail: licosa@licosa.com
http://www.licosa.com

NETHERLANDS/PAYS-BAS
De Lindeboom Internationale Publikaties
PO Box 202, MA de Ruyterstraat 20 A
NL-7480 AE HAAKSBERGEN
Tel.: (31) 53 574 0004
Fax: (31) 53 572 9296
E-mail: lindeboo@worldonline.nl
http://home-1-worldonline.nl/~lindeboo/

NORWAY/NORVÈGE
Akademika, A/S Universitetsbokhandel
PO Box 84, Blindern
N-0314 OSLO
Tel.: (47) 22 85 30 30
Fax: (47) 23 12 24 20

POLAND/POLOGNE
Głowna Księgarnia Naukowa
im. B. Prusa
Krakowskie Przedmiescie 7
PL-00-068 WARSZAWA
Tel.: (48) 29 22 66
Fax: (48) 22 26 64 49
E-mail: inter@internews.com.pl
http://www.internews.com.pl

PORTUGAL
Livraria Portugal
Rua do Carmo, 70
P-1200 LISBOA
Tel.: (351) 13 47 49 82
Fax: (351) 13 47 02 64
E-mail: liv.portugal@mail.telepac.pt

SPAIN/ESPAGNE
Mundi-Prensa Libros SA
Castelló 37
E-28001 MADRID
Tel.: (34) 914 36 37 00
Fax: (34) 915 75 39 98
E-mail: libreria@mundiprensa.es
http://www.mundiprensa.com

SWITZERLAND/SUISSE
BERSY
Route d'Uvrier 15
CH-1958 LIVRIER/SION
Tel.: (41) 27 203 73 30
Fax: (41) 27 203 73 32
E-mail: bersy@freesurf.ch

UNITED KINGDOM/ROYAUME-UNI
TSO (formerly HMSO)
51 Nine Elms Lane
GB-LONDON SW8 5DR
Tel.: (44) 207 873 8372
Fax: (44) 207 873 8200
E-mail: customer.services@theso.co.uk
http://www.the-stationery-office.co.uk
http://www.itsofficial.net

Adeco – Van Diermen
Chemin du Lacuez 41
CH-1807 BLONAY
Tel.: (41) 21 943 26 73
Fax: (41) 21 943 36 06
E-mail: mvandier@worldcom.ch

**UNITED STATES and CANADA/
ÉTATS-UNIS et CANADA**
Manhattan Publishing Company
468 Albany Post Road, PO Box 850
CROTON-ON-HUDSON,
NY 10520, USA
Tel.: (1) 914 271 5194
Fax: (1) 914 271 5856
E-mail: Info@manhattanpublishing.com
http://www.manhattanpublishing.com

STRASBOURG
Librairie Kléber
Palais de l'Europe
F-67075 STRASBOURG Cedex
Fax: (33) 03 88 52 91 21

Council of Europe Publishing/Editions du Conseil de l'Europe
F-67075 Strasbourg Cedex
Tel.: (33) 03 88 41 25 81 – Fax: (33) 03 88 41 39 10 – E-mail: publishing@coe.int – Web site: http://book.coe.int

Sales agents for publications of the Council of Europe
Agents de vente des publications du Conseil de l'Europe